Energy-based vegetation mapping

A case study in statistical quantum ecology

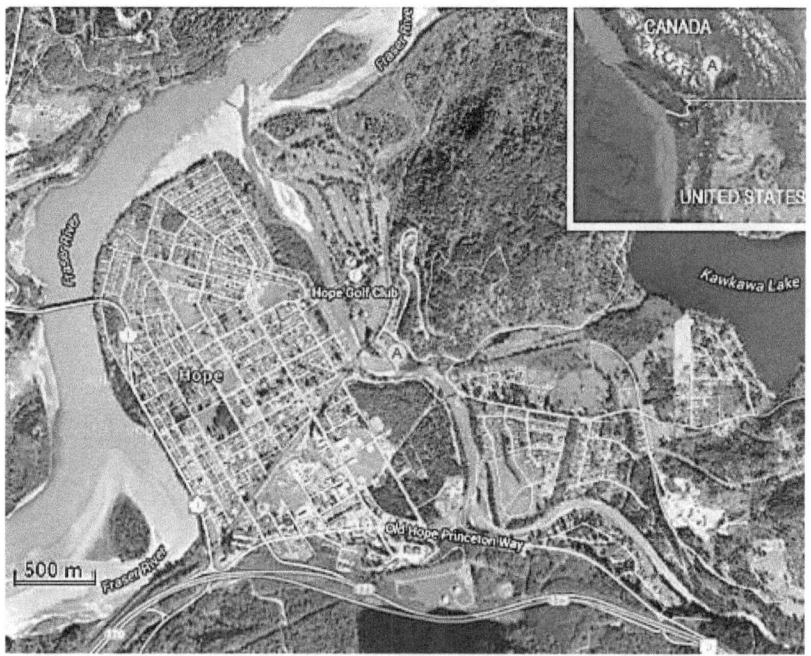

Map 1. Google map of Hope, British Columbia. The study site (49.381744 N, 121.42784 W) is located on the historic Coquihalla River floodplain within the triangular forest patch marked by A in the map. A large portion of the Hope's lower residential area is nested on Coquihalla alluvium between the Fraser River on the west and the Coquihalla River on the east.

Quantum Ecology provides the tools needed to map the vegetation's energy structure onto a ground grid. The mapping parameters are components of the energy structure issuing from historic phylogeny, current environmental forcing, and ubiquitous chance events. Concomitant instability maps reveal increased energy structural stability under decreasing species richness.

Dedicated to the memory of Professor Vladimir J. Krajina
my Ph.D. mentor in forest ecology at U.B.C.

ENERGY-BASED VEGETATION MAPPING

A case study in statistical quantum ecology

László Orlóci FRSC, FHAS e.m.

Ecologia Quantitativa, UFRGS, Porto Alegre, Brazil

With the technical assistance of
Márta Mihály
forest engineer

SCADA Publishing – Canada

Refer to this monograph as:

Orlóci, L. 2015. Energy-based vegetation mapping. A case study in statistical quantum ecology. SCADA Publishing, Canada. Online Edition: https://createspace.com/5495773

Look for these monographs:

Orlóci, L. 2014. The vegetation process. A holistic study of long-term community energetics in East Beringia. SCADA Publishing, Canada. Online Edition: https://createspace.com/4760258

Orlóci, L. 2013. Quantum analysis of primary succession. The energy structure of a vegetation chronosere in Hawai'i Volcanoes National Park. SCADA Publishing, Canada. Online Edition: https://createspace.com/4452597

Orlóci, L. 2013. Quantum Ecology. Energy structure and its analysis. SCADA Publishing, Canada. Online Edition: https://createspace.com/4406077

Orlóci, L. 2013. On the Energy Structure of Natural vegetation. In search for community governance rules. SCADA Publishing, Canada. Enlarged Online Edition: https://createspace.com/4153484

Orlóci, L. 2012. Self-organisation and Mediated Transience in Plant Communities. SCADA Publishing, Canada. Enlarged Online Edition: https://createspace.com/35 85127

Orlóci, L. 2012. Statistical Ecology. The quantitative exploration of nature to reveal the unexpected. SCADA Publishing, Canada. Online Edition: https://createspace.com/3476529

Orlóci, L. 2012. Statistical multiscaling in dynamic ecology. Probing the long-term vegetation process for patterns of parameter oscillations. SCADA Publishing, Canada. Online Edition: https://createspace.com/3830594

Orlóci. L. 2011. Problem flexible computing in statistical ecology. SCADA Publishing, Canada. Online Edition: https://createspace.com/3574792

```
Title ID: 5495773
ISBN-13: 978-1512180763
ISBN-10: 1512180769
```

V 2015-05-12

Find further information at URL
https://sites.google.com/site/statisticalecology/

Please send all communications to: lorloci@uwo.ca

Energy-based vegetation mapping

Contents

Preface

This is the author's fifth monograph in the series on the theory and modus operandi of *Quantum Ecology*. The subject is presented at this time wrapped in the problem domain of vegetation mapping. Considering that mapping of our medium involves stand-level vegetation units, quantum ecology's holistic approach - which puts the stand's potential energy structural state into focus as a classificatory criterion - is particularly well suited for the task. This is radically different from the traditional practice of vegetation energetics which targets population level calorific flow.

The theoretical significance of quantum ecology's stand-level approach resides with the redirection of effort in energetic studies to potential energy. Implementation in practice is made simple by the *resonator complex* model borrowed from Max Planck's quantum theory. The model translates smoothly into phytosociology terms: *complex* becomes the *unit plant community* or *stand,* and *resonator* a *plant population* (species, multispecies, etc.). The universal scalar for the energy level, or more correctly the potential energy level of any resonator complex is Max Planck's energy-based entropy $E = - \ln P$.

This is a suitable point to respond to the frequently expressed misconception that $E = - \ln P$ as an *alternative energy parameter* renders the use of the term "energy" a metaphor. My response goes like this:

1. To me the term *alternative* conjures the fact that energy has never been defined, therefor it cannot be measured directly, only indirectly by its manifestations. Ergo there must be as many possible alternative energy parameters as there are possible manifestations which are measurable.

2. In a specific case calorie is the alternative parameter for energy. For example, starting with an amount of air free water, the energy needed to raise its temperature from 287.65 to 288.65 $^{\circ}$K is Cal = 4.1855m J. In this m is the amount of water in grams and J stands for Joule's scale.

3. Function $E = -\ln P$ is an alternative parameter for the energy sate of a resonator complex. In this case, energy is measured in natural units (nat).

These are two alternative parameters of energy among as many proxy parameters as there are possible manifestations of spent energy.

Max Planck's equation is parameterised by the actual number of resonators (n) and the sum of the energy units (T) of the resonators in the complex. We link T to plant population performance. As for the definition of P we have,

$$P = \frac{1}{C} \text{ and } C = \frac{(T+n-1)!}{T!(n-1)!} \gg \frac{(T+n)^{T+n}}{T^T n^n}$$

The quantum theoretical interpretation of E requires that specific regularity conditions are fulfilled:

1. The number of energy units, the quanta, is a manifestations of spent energy and not the amount of energy itself. In this sense, the quantity of a plant population is an energy unit count, an integer.

2. The rule of chance in the resonator complexes assembly must be assumed. In a vegetation survey which use sample plot based relevés this condition is fulfilled if the sample plot is homogeneous. Homogeneity implies that the ground pattern of the vegetation or environment is random.

Quantum ecology defines the vegetation stand's energy state (E) as a sum of three independent potential energy components issuing from sources in the manner of the energy equation:

$$E = E_{Phy} + E_{Env} + E_{Rnd}$$

The first two terms on the right hand side of the equation are measurable. Among these E_{Phy} is proxy for the amount of energy spent in the course of historic phylogeny for the local flora to reach its current level of diversity. The concomitant E_{Env} is an estimate of energy spent in the course of current environmental mediation on the plant populations' assembly into distinct vegetation stands. The third term, E_{Env}, is a residual in the manner of the difference $E_{Rnd} = E - E_{Phy} - E_{Env}$. As a provisional decision, not uncommon in statistical ecology's practice, we assign the unidentified processes responsible for residuals to the grab-bag of the so called errors generated by *random events*. An "error" designation notwithstanding, further analysis of the residuals often reveals the presence of determinism, the opposite of randomness, pointing up the need for model adjustments.

Considering the novelty of *quantum ecology,* it will serve well its understanding of the approach if we place it into perspective regarding the proximity to *statistics*. Our first approximation for statistics definition is rather pragmatic. To us it is nothing more in principle than a coupling of probability theory with some characteristic functionality of physical science. We refer to such a coupling as a "statistical dialect". Each possible dialect is unique. For example, when probability theory is coupled with the *system of moments and product moments*, foundation is laid for the "Fisherian" statistical dialect. This is the dialect most commonly applied by biologists. When the coupling involves probability theory and Rényi's information divergence of order one, the foundations of Kullback's statistical dialect is defined. There is no limit to the number of statistical dialects. Accordingly, we consider quantum ecology as one of the statistical dialects in which probability theory is coupled by quantum theory by way of Max Planck's energy-based entropy function. The basic question perplexing the users of any statistical dialects is the same: to what extent the observed state of a given complex can be considered a chance induced event. After this the approaches diverge. Quantum ecology answers this question in energy terms.

The inception of coupling energy-based entropy with probability theory as a statistical tool in ecology came quite naturally to me when the need for it presented itself. Students' questions brought me onto it. The project in question called for the serial measurement of stand-level energy in field studies. They found themselves not wiser by what they got out from accounts of studies which targeted calorific flow in the food chain. Techniques were needed by which community energetics can be studied based on the phytosociologist's usual kind of survey type vegetation data.

The problem struck me as one which should be concerned with scaling the potential energy state in stands. The solution took shape in a rather protracted on-and-off kind of process from first inception of the problem as the Brazilian CNPq's visiting professor in Dr. Valério DE Patta Pillar's quantitative ecology lab at UFRGS in Porto Alegre,. The project started to gain momentum by chance after I came across Max Planck's 1901 article "On the law of distribution of energy in the normal spectrum" the second time, roughly six decades after having had it as compulsory reading during my engineering studies. I rediscovered it after perusing through Stephen Hawking's "The dreams that stuff is made of". It was Márta's Christmas present to me in 2012. Solution of the energy scaling problem in hands, quantum statistics could be developed for the analysis of holistic energetics. I called it *Quantum Ecology* (Orlóci 2013a, b, c, 2014). I set as quantum ecology's first order objective the isolation of components in the energy structure of real vegetation stands in Nature in the manner of the energy equation $E = E_{Phy} + E_{Env} + E_{Rnd}$.

This Essay presents numerical examples as its predecessors did, but at this time the centre stage is assigned to stand-level energy mapping on the Coquihalla floodplain. The site is identified by "A" on the Google maps (page 1). The forest vegetation of the site has been surveyed by M. Mihály and team in 1976. Beyond being the foundation of the examples in this book, the 1976 record set has gained considerable historic significance. It fix conditions four decades back in time, which no longer exist in the site.

On our first visit to the Coquihalla floodplain in 1976, we were surprised by the natural high forest in the site, mature, second growth at that time. Sharply defined, concomitant patterns of forest types and the natural floodplain topography were most obvious. We saw for the site a future as a nature reserve or a least an open teaching laboratory for floodplain ecology. I wrote about this to Professor Vladimir J. Krajina[1], my Ph.D. mentor at the University of British Columbia, who was deeply involved at that time in his career-crowning Ecological Reserves Program[2]. I suggested to him a visit to the Coquihalla site. Dr. Krajina did visit the site and responded emphatically: "Wie es im Buche steht"!

V.J. Krajina's interest fell short of success to gain ecological reserve status for the site. I salute his memory for the effort and, beyond that, for the futuristic ecological views he taught in class and implanted by personal deeds into the forestry profession. V.J. Krajina's ideas have developed deep roots in his adopted country's forest management.

Corrections and revisions suggested by Professor Mathew M. Mukkattu are gratefully acknowledged.

L. Orlóci

London, 2015 May 1

---

[1] http://ecoqua.ecologia.ufrgs.br/%7Elorloci/Koa/1001%20ZIVA.pdf
[2] http://ecoreserves.bc.ca/2012/03/12/contributions-of-vladimir-krajina-to-ecological-reserves-in-bc/

Coquihalla survey site

Virtual tour

At the time of the survey in 1976, a native high forest covered the site.[3] Some exceptional specimens of Douglas fir *(Pseudotsuga menziesii)* and Red-cedar *(Thuja plicata)* toped 70 meters in height at age 80 years. The site received regular overflow at flood stage from the Coquihalla River. The site was designate as a municipal park. Camping was allowed in one section of the forest west of our belt transect and south on the opposite side of Kawkawa Lake Road (Map 2).

Closer examination of the maps reveals substantial engineering and land use activities following 1976. These were of the kind one would expect in preparation for large scale expansion of the municipality's residential area onto the floodplain. If in deed housing extension was an objective, the project must have come

---

[3] At this point we should make the linkage of potential biomass production to the types to emphasize the importance of bench height. The measured height of dominant Douglas-fir *(Pseudotsuga menziesii)* is an indication:

Type	Polystichum	Mahonia	Gaultheria
Bench height	Low	Medium	High
Sample size	13	12	10
Mean tree height m	53.88515	45.75046	28.05028
STD m	2.484593	2.261358	6.326595

These are 1976 records. Age range 60 to 80 years.

to a rather abrupt halt after construction of a single row of family homes right under the dyke on the River Promenade (Map 2). Incidentally, this street crosses the lowest lying section of the historic floodplain.

The relative height of the dyke can be judged on Map 3 in cpparison to the size of the vehicle parked in clear view on the Promenade. The dykes' actual height matches the theoretical 200-year flood level in the Coquihalla River.

Map 2. The survey site with the belt transect trace from marker "A". Note: Pon this map, the transect positions corrects an earlier map of the same site.

Map 3. Google's street level view of the River Parade, looking west, just north of marker "A" on Maps 1 and 2. High dyke on the right side protects a single row of family homes (left). The height of the dyke matches the 200-year flood level, established by the 1985 hydraulic survey.

Topography

Map 4 displays the macro topography of the Coquihalla River's lower course from the point where it leaves the canyon downstream to its confluence with the Fraser River.

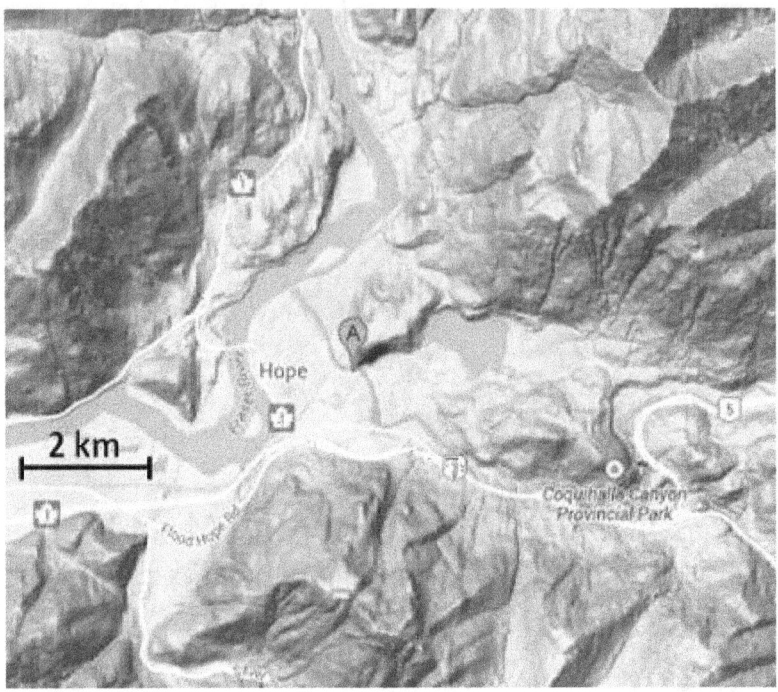

Map 4. General topography of the Hope region by Google.

The local topography of the survey site itself is typical of a well-developed floodplain. The surface is fragmented by distinct levels, which we call *benches*. These are flat surfaces, slightly inclined away from the river, each sensitively sculpted by flood-related natural aggradation and erosion. Bench height in the site varies from 1.3 m (**B1**), to 1.6 m (**B2**) and 3.2 m (**B3**) above the water level in the river on the site at 10 am on the first day of sampling, July 7, 1976.

River hydraulics 1976

Figures 1 and 2 display the historic 200-year floodplain limits, the 200-year flood level contours, and other land survey details. It is seen that the elevation maximum on the transect is about 53 m and minimum about 50 m above the zero mark on the Fraser

River's water level gauge (Hope bridge, Map 1). The 50 m drop materializes over close to 2000 areal meters. Translated into degrees, the incline upstream is about 0.14^O.

Figure 1a. Topographic map of the site. The 200-year flood level contours are shown within the 200-year floodplain marked by the shaded, heavy dark line. Source: B.C. Ministry of Environment, Water Management Branch base map.

The flow in the river is expected near $2m^3/s$ from October to March, but up to $12m^3/s$ in May and June. The flow can be torrential with a 24hr lag after warm days or following heavy rain on high elevations in the Cascades over the 740 km^2 watershed. Highest peak in the watershed is nearly 2000 m above sea level.

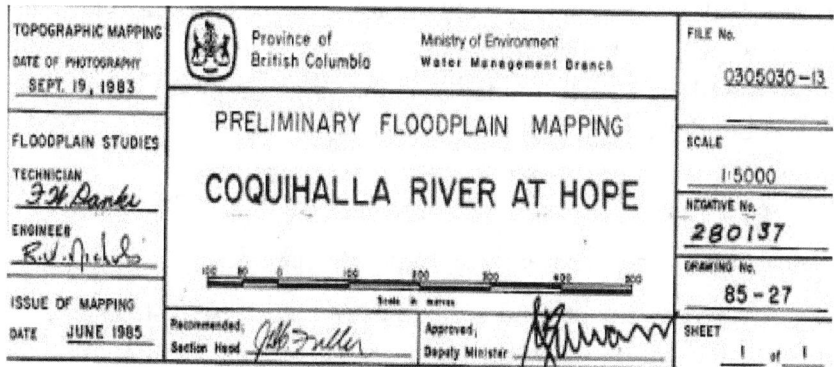

Figure 1b. Specifics of topographic mapping prior to the construction of the present dyke system. Topographic survey was completed in 1983 and map issued in 1985.

Sampling design and C/A scale

Sample plots of two sizes (10x10 and 20x20) were laid at the nodes of a randomly sited systematic grid as drafted below:

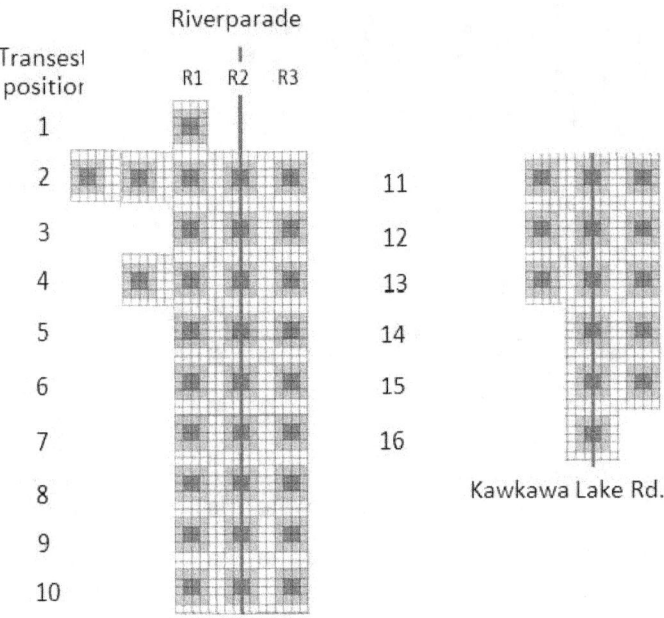

Figure 2. Minimum grid resolution is 5m x 5m. The distance of nodes is 30m. The central vertical line in the figure is the trace of the transect (Map 2). The total number of plots is 45. The size of the largest rectangular block of nodes which could be fitted on to the grid for mapping purposes is 12 x 3 = 36. At each node the inner sample plot (10 m. sq.) is for the lesser vegetation, the outer (20 m. sq.) for trees.

Plant population performance are measured on phytosociology's cove/abundance (C/A) scale:

Alphanumeric values	Rare*	Spo-radic**	1-5%	5-25%	25-50%	50-75%	75-100%
Numeric scale	1	2	3	5	7	8	9

*solitary specimens, no cover value; **sporadic specimens, low cover value

Cover refers to the ground surface estimated in five states. Abundance implies plentifulness. This has two states on our version of the scale (rare, sporadic) which split the first category of Braun-Blanquet's 1928 abundance scale in two. We interpret the integer values of the numeric scale as energy unit counts from 1 to 9.

Coquihalla data set

The total number of plant species identified within the 45 sample plots is 72. We give the data for the 40 leading taxa (38 species plus two shrub forms of trees) in Tables 1 and 2. "Leading" taxa are those which have total cover value attained in the 45 plots not less than 8. The low shrub form of *Thuja plicata* is at this limit.

Table 1. Systematic code for species. The L key indicates hierarchical level.

#	L0 - Species	L1 - Class or higher	Code	L2 - Order	Code
1	*Thuja plicata (shrub)*	Pinopsida	6	Pinales	13
2	*Holodiscus discolor*	Magnoliopsida	4	Rosales	15
3	*Mnium spinulosum*	Bryopsida	2	Eubryales	9
4	*Lonicera ciliata*	Asterids	1	Dipsacales	7
5	*Claytonia sibirica*	Magnoliopsida	4	Caryophyllales	4
6	*Rosa gymnocarpa*	Magnoliopsida	4	Rosales	15
7	*Spiraea douglasii*	Magnoliopsida	4	Rosales	15
8	*Amelanchier florida*	Magnoliopsida	4	Rosales	15
9	*Vaccinium parviflorum*	Asterids	1	Ericales	8
10	*Acer macrophyllum*	Eudicots	3	Sapindales	16
11	*Galium triflorum*	Asterids	1	Gentianales	10
12	*Acer macrophyllum (shrub)*	Eudicots	3	Sapindales	16
13	*Osmorhiza chilensis*	Asterids	1	Apiales	1
14	*Circaea alpina*	Magnoliopsida	4	Myrtales	12
15	*Pteridium aquilinum*	Pteridopsida	7	Dennstaedtiales	6
16	*Vaccinium membranaceum*	Asterids	1	Ericales	8
17	*Smilacina stellata*	Monocots	5	Asparagales	2
18	*Tsuga heterophylla*	Pinopsida	6	Pinales	13
19	*Linnaea borealis*	Asterids	1	Dipsacales	7
20	*Chimaphila umbellata*	Asterids	1	Ericales	8
21	*Goodyera oblongifolia*	Monocots	5	Asparagales	2
22	*Dicentra formosa*	Magnoliopsida	4	Ranunculales	14
23	*Trientalis latifolia*	Eudicots	3	Ericales	8

24	*Rubus spectabilis*	Magnoliopsida	4	Rosales	15
25	*Clintonia uniflora*	Monocots	5	Liliales	11
26	*Mnium insigne*	Bryopsida	2	Eubryaales	9
27	*Symphoricarpos albus*	Asterids	1	Dipsacales	7
28	*Lactuca canadensis*	Asterids	1	Asterales	3
29	*Rhytidiadelphus loreus*	Bryopsida	2	Eubryales	9
30	*Disporum hookerii*	Monocots	5	Liliales	11
31	*Gaultheria shallon*	Eudicots	3	Ericales	8
32	*Polystichum munitum*	Pteridopsida	7	Dennstaedtiales	6
33	*Pachistima myrsinites*	Eudicots	3	Celastrales	5
34	*Achlys triphylla*	Magnoliopsida	4	Ranunculales	14
35	*Eurhynchium oreganum*	Bryopsida	2	Eubryales	9
36	*Mahonia nervosa*	Magnoliopsida	4	Ranunculales	14
37	*Thuja plicata*	Pinopsida	6	Pinales	13
38	*Hylocomium splendens*	Bryopsida	2	Eubryales	9
39	*Acer circinatum*	Eudicots	3	Sapindales	16
40	*Pseudotsuga menziesii*	Pinopsida	6	Pinales	13

Table 1 continued

#	L3 - Family	Code	L4 - Genus	Code
1	Cupressaceae	15	Thuja	32
2	Rosaceae	3	Holodiscus	12
3	Mniaceae	9	Mnium	20
4	Caprifoliaceae	18	Lonicera	18
5	Portulacaceae	5	Claytonia	6
6	Rosaceae	3	Rosa	27
7	Rosaceae	3	Spiraea	30
8	Rosaceae	3	Amelanchier	3
9	Ericaceae	13	Vaccinium	34
10	Sapindaceae	1	Acer	1
11	Rubiaceae	2	Galium	11
12	Sapindaceae	1	Acer	1
13	Apiaceae	21	Osmorhiza	21
14	Onagraceae	7	Circaea	5
15	Dennstaedtiaceae	14	Pteridium	25
16	Ericaceae	13	Vaccinium	34
17	Ruscaceae	21	Smilacina	29
18	Pinaceae	6	Tsuga	14
19	Caprifoliaceae	18	Linnaea	17
20	Pyrolaceae	4	Chimaphylla	4
21	Orchidaceae	7	Goodyera	13
22	Fumariaceae	12	Dicentra	8
23	Myrsinaceae	8	Trientalis	33
24	Rosaceae	3	Rubus	28
25	Liliaceae	10	Clintonia	7
26	Mniaceae	9	Mnium	20
27	Caprifoliaceae	18	Symphoricarpos	31
28	Asteraceae	20	Lactuca	16
29	Hypnaceae	11	Rhytidiadelphus	26
30	Colchicaceae	16	Disporum	9
31	Ericaceae	13	Gaultheria	12
32	Dennstaedtiaceae	14	Polystichum	23
33	Celastraceae	17	Pachistima	22

34	Berberidaceae	19	Achlys	2
35	Hypnaceae	11	Eurhynchium	10
36	Berberidaceae	19	Mahonia	19
37	Cupressaceae	15	Thuja	32
38	Hypnaceae	11	Hylocomium	15
39	Sapindaceae	1	Acer	1
40	Pinaceae	6	Pseudotsuga	24

Table 2. C/A totals for the three floodplain benches (B1, B2, B3) and for the pooled sample (B4). In all, 40 leading taxa are included and 45 sample plots (14 plots on bench B1, 20 on bench B2, and 11 on B3). The code values, copied from Table 1 for quick review, have significance only as unordered labels by which we impose a proxy phylogenetic structure on the sample.

#	L0 - Species	Taxonomic map				Total C/A			
		L1	L2	L3	L4	B1	B2	B3	B4
1	*Thuja plicata (shrub)*	6	13	15	32	0	0	8	8
2	*Holodiscus discolor*	4	15	3	12	0	0	13	13
3	*Mnium spinulosum*	2	9	9	20	15	2	0	17
4	*Lonicera ciliata*	1	7	18	18	0	0	21	21
5	*Claytonia sibirica*	4	4	5	6	0	5	17	22
6	*Rosa gymnocarpa*	4	15	3	27	0	0	23	23
7	*Spiraea douglasii*	4	15	3	30	2	8	15	25
8	*Amelanchier florida*	4	15	3	3	26	0	0	26
9	*Vaccinium parviflorum*	1	8	13	34	3	8	16	27
10	*Acer macrophyllum*	3	16	1	1	30	0	0	30
11	*Galium triflorum*	1	10	2	11	23	3	5	31
12	*Acer macroph. (shrub)*	3	16	1	1	26	6	0	32
13	*Osmorhiza chilensis*	1	1	21	21	0	5	30	35
14	*Circaea alpina*	4	12	7	5	0	2	33	35
15	*Pteridium aquilinum*	7	6	14	25	25	10	0	35
16	*Vaccinium membranaceum*	1	8	13	34	18	18	0	36
17	*Smilacina stellata*	5	2	21	29	0	4	34	38
18	*Tsuga heterophylla*	6	13	6	14	32	14	0	46
19	*Linnaea borealis*	1	7	18	17	47	5	0	52
20	*Chimaphila umbellata*	1	8	4	4	0	39	15	54
21	*Goodyera oblongifolia*	5	2	7	13	0	54	0	54
22	*Dicentra formosa*	4	14	12	8	0	61	0	61
23	*Trientalis latifolia*	3	8	8	33	7	38	19	64
24	*Rubus spectabilis*	4	15	3	28	8	43	16	67
25	*Clintonia uniflora*	5	11	10	7	0	82	3	85
26	*Mnium insigne*	2	9	9	20	69	17	2	88
27	*Symphoricarpos albus*	1	7	18	31	70	18	4	92
28	*Lactuca canadensis*	1	3	20	16	25	47	34	106
29	*Rhytidiadelphus loreus*	2	9	11	26	55	54	0	109
30	*Disporum hookerii*	5	11	16	9	69	50	0	119
31	*Gaultheria shallon*	3	8	13	12	0	43	82	125
32	*Polystichum munitum*	7	6	14	23	4	65	72	141
33	*Pachistima myrsinites*	3	5	17	22	105	47	1	153
34	*Achlys triphylla*	4	14	19	2	63	94	17	174
35	*Eurhynchium oreganum*	2	9	11	10	42	99	51	192
36	*Mahonia nervosa*	4	14	19	19	37	132	43	212
37	*Thuja plicata*	6	13	15	32	38	122	52	212
38	*Hylocomium splendens*	2	9	11	15	23	123	73	219
39	*Acer circinaum*	3	16	1	1	106	105	29	240

40	*Pseudotsuga menziesii*	6	13	6	24		113	121	88	322
					n		27	34	28	40
					T		1081	1544	816	3441

We can identify two kinds of components in each data element by the questions they attract: what kind and how much? We chose to understand "kind", the qualitative component, in phylogenetic terms since we deal with plant species. It will be seen that the phylogeny related component can be measured analytically on the basis of the proxy phylogenetic tree (dendrogram) which the systematic code defines (Table 1). But for this, we need to attach C/A cumulants to the nodes of the tree (Orlóci 2013a). By choice our proxy tree has 5 levels, L1 to L4 for Class, Order, Family and Genus, and L0 for species.

The component which attracts the question "how much" or "how abundant" is the quantitative component. There are C/A totals entered for these in Tables 2 for 45 sample plots and in Table 3 for 36 sample plots.

Table 3. Each entry in the table resents the C/A total T and corresponding species number n a single plot at the corresponding position of the sampling grid (see previous section). Column R4 contains row totals. "Position" number increases away from marker "A" (Map 2).

Position		T matrix				n matrix			
global	block	R1	R2	R3	R4	R1	R2	R3	R4
2	1	82	82	73	237	17	17	15	49
3	2	79	71	70	220	17	15	17	49
4	3	83	87	80	250	20	21	18	59
5	4	70	67	88	225	17	17	18	52
6	5	81	89	62	232	23	21	17	61
7	6	71	79	81	231	18	17	21	56
8	7	88	71	73	232	20	14	17	51
9	8	80	75	88	243	19	19	23	61
10	9	74	73	77	224	20	17	21	58
11	10	86	80	76	242	22	18	17	57
12	11	64	73	81	218	18	18	19	55
13	12	76	79	75	230	18	17	18	53

Potential energy structure

The resonator complex

We already defined energy based vegetation classification as an exercise in the application of energy-based entropy (Planck 1901). With his we can scale the potential energy level in the vegetation stand of each sample plot in natural units (nats).

We present the topic on the following pages in the context of quantum theory's *resonator complex.* Quantum ecology assigns the role of the complex to the unit plant community which is a *homogeneous vegetation stand within area limits.* The role of the resonators goes to the stand's component *plant populations.* We call a plant population *species* when its recognition is inheritance based. Populations of another types are identified by their *functional types.* From a purely mechanistic point of view, species and functional types are *character set types* which differ by the set of traits on the basis of which they are recognised as sequentially or hierarchically arranged entities (Orlóci, 1991).

It is re-emphasized that a vegetation stand must satisfy the regularity conditions of homogeneity to qualify as a resonator complex. This condition is satisfied when the ground patterns of vegetation and environment are random within the sample plot.

To see clearly the points to be made we start with the symbolic record of a resonator complex **X**:

Res.	1	2	3	4	5	6	7	8	9	10
X	8	12	10	5	12	3	22	9	11	8
P	0.08	0.12	0.10	0.05	0.12	0.03	0.22	0.09	0.11	0.08
Q	0.1	0.1	0.1	0.1	0.1	0.1	0.1	0.1	0.1	0.1

To us, the resonators (1 to 10) represent distinct plant populations, species or otherwise, generally called *taxa*. The X quantities are energy unit counts in the guise of the number of individual plants per taxa within a sample plot, cover-abundance estimates per taxa, frequencies, etc. Such a data set as **X** is a common starting point in survey-based vegetation studies.

We consider a common dichotomy in the above regard. One branch of the dichotomy leads to *diversity analysis* and the other to *holistic energetics*. Diversity analysis is concerned with *disorder* in **X**. Holistic energetics is concerned with the *potential energy level*.

Disorder

Information theory is our guide in diversity analysis. The choice of scaling functions answers to specific questions, such as:

Q1. How intense is the disorder in X as a distribution? Since the answer is scale-dependent, Rényi's (1961) entropy of order α is a sensible choice,

$$H_\alpha = \frac{1}{1-\alpha} \sum_{i=1}^{n} p_i^{\alpha}$$

In this α is the scale variable and n is the number of taxa. When α is equal to zero, H is a measure of species richness. When the value of α is approaching the limit one, H is measuring disorder in the vegetation stand to which ecologists usually refer as Shannon's (1948) diversity index. Other higher order H define multiscale disorder in **X** with or without identified utility in ecological diversity studies. The regularity condition to be satisfied at any order is that that the n individual p values add up to exactly one.

Q2. How far the observed disorder departs from one that is specified by theory of observation, fixed by the vector elements

$(q_1 \ q_2 \ ... \ q_n)$? The ideal choice to measure the divergence is Rényi's (1961) information of order α,

$$I_\alpha = \frac{1}{\alpha-1} \sum_{i=1}^{n} \frac{p_i^\alpha}{q_i^{\alpha-1}}$$

At order one, 2I is Kullback's (1959) *minimum discrimination information statistics*. Kullback shows that the sampling distribution of 2I is asymptotically Chi-squared with specified degrees of freedom, under specified restrictive assumption.

For the time it is sufficient to point out that diversity analysis operates at the resonator level **p** and **q**. This makes diversity analysis focus on disorder in **X**. This differs sharply from quantum analysis which we apply to the same type of survey data on the complexes level.

Potential energy

It is not an objective of this book to re-explain the basics in detail or to develop new, step-by-step training exercises. The reader finds much on the subject in my CraeteSpace books[4]. The energy parameter about which the study of the potential energy state of a vegetation stand revolves is Max Planck's

$$S_N = k \ln W + contant$$

The symbols: ln logarithm to base **e**, **k** universal constant, **W** a probability, **constant** amount of translation to desired origin.

---

[4] Orlóci, L. 2014. The vegetation process. A holistic study of long-term community energetics in East Beringia. SCADA Publishing, Canada. Online Edition: https://createspace.com/4760258

Orlóci, L. 2013. Quantum analysis of primary succession. The energy structure of a vegetation chronosere in Hawai'i Volcanoes National Park. SCADA Publishing, Canada. Online Edition: https://createspace.com/4452597

Orlóci, L. 2013. Quantum Ecology. Energy structure and its analysis. SCADA Publishing, Canada. Online Edition: https://createspace.com/4406077

Orlóci, L. 2013. On the Energy Structure of Natural vegetation. In search for community governance rules. SCADA Publishing, Canada. Enlarged Online Edition: https://createspace.com/4153484

We rewrite the core expression by different symbols and simplified, such as

$$E = \ln\frac{1}{P} = \ln C$$

By setting k equal to -1 we do not affect the comparability of individual E values between stands. Leaving out the "constant" means returning to the natural origin.

The symbol P needs explanation. It is the same as Max Planck's W, the probability of an exact duplicate of the observed vegetation complex to assemble under the rule of pure chance under already stated homogeneity conditions. Accordingly, we have

$$P = \frac{1}{C} \text{ in which } C = \frac{(T+n-1)!}{T!(n-1)!} \approx \frac{(T+n)^{T+n}}{T^T n^n}$$

The value of C indicates the number of possible complexes that can be assembled from any n resonators with aggregated complex level total energy unit count $T = \sum_{i=1}^{n} X_i$. It is obvious from $C \times P = 1$ that each complex is considered to be an equal probability event. We refer to C-1 as the number of possible *ghost complexes*.

The basic potential energy model of the Coquihalla case study is

$$E = E_{Phy} + E_{Env} + E_{Rnd}$$

This is quantum ecology's energy equation by which the energy structure of vegetation stand is described. According to this, the energy structure has three additive components: spent energy issuing from historic phylogeny (Phy) which have created the species, spent energy issuing from current environmental forcing (Env) which sorts species into distinct vegetation stands, and spent energy associated with purely random events (Rnd).

Some properties of E

Consider the simple equation, $-\ln\frac{1}{e} = 1$. This is telling us that potential energy when measured by E is in natural units or nats.

The term "nat" was used earlier by R.P. McIntosh (1967) and later adopted by E.C. Pielou (1977) to designate the unit in which disorder based diversity is expressed.

Another self-evident property is the consequence of parameterisation of E by n and T. This is the exact reason why we say that E is a holistic parameter of the energy level in the plant community complex. The amount of disorder in **X** on the resonator level has no effect on E.

Yet another property is the unique linkage of E to *potential energy*. This distances E from energy parameters linked to energy transported in *calorific flow* through trophic networks of populations.

The resonators of a complex are scaled by energy unit counts. When the energy content of one energy unit is given as ε, the total energy in the n resonator complex is T x ε. Considering the fact that the energy units, the quanta, are indivisible, T has to be an integer number and it cannot be less than n.

Ecological identity of E

In its broader consequences the exercise of vegetation analysis based on E is equivalent to redefining a new, holistic paradigm for the study of vegetation energetics. I refer to this paradigm as *Quantum Ecology*. When E is applied to survey-type vegetation data, Quantum Ecology moves outside the strictly theoretical framework established for resonator complexes by Max Planck (1901). This is not without penalties and we have to be prepared to accept that E becomes context dependent. E still retains validity operationally, but the context limits comparability. The context has to be identified and stated. In the simplest case, it is sufficient to identify the sampling design and data type.

We emphasise that E for the phytosociologist is an alternative parameter for scaling the potential energy in vegetation stands. Because of this, an E-based analysis generates results on the stand level. This contrasts sharply with disorder based diversity analysis which works with sum of n "p ln p", such as in Shannon's

(1948) entropy function. To make the consequences clear it has to be pointed out that whereas parametrisation of p ln p is on the level of the resonator probabilities p_1, p_2, ..., p_n, the parameterization of P in $E = -\ln P$ does not involve the resonator probabilities. This is clear from what we see in the working equation,

$$E = nH = -\ln P = \ln\frac{(T+n-1)!}{T!(n-1)!} \approx \ln\frac{(T+n)^{T+n}}{T^T n^n}$$

When we introduce the new symbol nH as an alternative to E, it emphasizes that E is a scalar of total potential energy. But nH may be read also as n time H in which case H is the one-resonator (average) energy level $H = \dfrac{E}{n}$ in the complex.

For reasons of completeness further comparisons are in order. Students should see the sharp distinction of $H = \dfrac{E}{n}$, which represents the one-resonator en-ergy-based entropy, and the disorder-based diversity scalars traditionally des-ignated by H. Energy-based entropy is Max Planck's quantum theoretical no-tion. Diversity analysis finds its theoretical foundations, but not necessarily its origins, in communications theory (Shannon 1948), information statistics (Kullback 1951), and particularly in the information theoretical work of Rényi (1961). The qualifier "particularly" may appear forced, yet appropriately used, considering that we can place almost any logarithmic species diversity meas-ure meaningfully under the umbrella of Rényi's generalized entropy H_α. To summarise what we have already said:

1. Shannon's entropy function $H = -\sum_{i=1}^{n} p_i \ln p_i$ is equivalent to Rényi's general-ised entropy $H_\alpha = \dfrac{1}{1-\alpha}\sum_{i=1}^{n} p_i^\alpha$ when the scale parameter α is approaching the limit value 1.

2. In Kullback's minimum information statistics $2I = 2\sum_{i=1}^{n} p_i \ln\dfrac{p_i}{q_i}$, the p terms are the same as in Shannon's H, and I is Rényi's generalised information $I_\alpha = \dfrac{1}{\alpha-1}\sum_{i=1}^{n} \dfrac{p_i^\alpha}{q_i^{\alpha-1}}$ when α is approaching the limit value 1. Letter q is symbol in

both cases for the random expectation of p. The source of q is theory or some empirical finding. Symbol n represents the number of taxa as before whose proportional quantity in the stand is $p_1, p_2, \ldots, p_n$. We note once more that for p_i we reach down into elements in the T-totalled and n-valued data vector **X** and define it as $p_i = \dfrac{X_i}{T}$. In sharp contrast, the scalar E is stand level and depends entirely on n and T.

3. $H = \dfrac{E}{n}$ is not the same as Whittaker's β diversity. It must be seen further that from a theoretical point of view the object's type on which entropy analysis is performed does not define entropy of a new generic type. Whittaker's β diversity is in deed just another case of Rényi's entropy of order one.

4. One more supplemental note: under specific regularity conditions Kullback's 2I is asymptotically distributed as a chi-squared variate with defined degrees of freedom. Parameter E is readily converted to a probability in the manner of $P = e^{-E}$ where e is the natural base.

Summary of equations written for E

Energy-based entropy level in the complex

$$E = nH = -\ln P = (T+n) \ln (T+n) - T \ln T - n \ln n$$

One resonator energy level,

$$H = \frac{E}{n}$$

Structural instability level,

$$w_{AB} = 1 - P_A^2 - P_B^2$$

The w_{AB} is measured in squared probability units

Definitions:

$P_A = e^{-E|A}$ -- E-based probability of complex A under the complete rule of chance. The one resonator version of this is $P_A = e^{-E/n|A}$.

$P_B = 1 - P_A$ -- probability of the ghost complex for A.

The values of w_{AB} range from 0 (complete stability) to 0.5 (complete instability). This can be easily seen if 1 or 0.5 is substituted for P_A. How do we interpret w_{AB} ? There is more than one way to do this. For example, as w_{AB} increases, the chance of complex A flipping into one of its ghost states B **by pure chance** increases. Since $\omega_{AB} = \sqrt{2w_{AB}}$ is a probability, the energy-based entropy discriminating against stability is

$$^{-}m\omega_{AB} = - \ln\,(1\text{-}\omega_{AB})\ \text{nats}$$

We call this the *unit instability moment.* This is the strength of instability or equivalently the unit linear moment for the vegetation complex's energy structure to flip into a ghost state. The strength of stability is then

$$m\omega_{AB} = - \ln\,(\omega_{AB})\ \text{nats}.$$

We call this the *unit stability moment.*S

Emergent (ghost) potential energy for vegetation seres
When A is a single complex or a sere of complexes is joined to sere B, a single complex or a sere of complexes different from A, the emergent energy-based entropy is

$$gnH(A) = dnH(A) - nH(A) \geq 0$$

In this,

$$dnH(A) = nH(A + B) - nH(B)$$

This is the energy quantity by which a spatial or temporal sere's total energy level is raised when sere B is extended to A+B. The energy imported is gnH + nH. This is valid for seres of complexes, provided that A and B are none overlapping and the following are true:

T(A) – grand total in A
T(A+B) -- grand total in A+B
n(A+B) – number of nonzero values in A+B
n(A) – number of taxa in A
nH(A+B) – potential energy level based on T=T(A+B) and n=n(A+B)
nH(B) – potential energy level based on T=T(A+B)-T(A) and n=n(A+B)-n(A)

One-complex emergent potential energy

$$dH(A) = \frac{dnH}{n(A)}$$

In this case the pivot on which two seres are joined is considered a member of sere A.

$$dH(B) = \frac{dnH}{n(B)}$$

This is the case when the pivot on which two seres join is a member of sere B.

Further essentials

It should be seen how the analytical focus is shifted in the above from calorific flow to the potential energy structure of the complex. We emphasise:

1. Our basic scalar $E = -\ln P$ is neither Alfréd Rényi's generalised entropy H_α nor his generalised information I_α.

2. In quantum analysis parameter P is a combinatorial function of T and n such as

$$C = \frac{(T+n-1)!}{T!(n-1)!} \approx \frac{(T+n)^{T+n}}{T^T n^n}$$

This is not the same as C in Brillouin's (1968) information equation $I = \log_2 C$.[5] Brillouin uses the combinatorial

$$C = \frac{N!}{X_1! \, X_2! \, \ldots \, X_s!}$$

Brillouin's X_i is an element of in an s-valued N totalled distributions, the type from which Rényi's $p_i = \frac{X_i}{N}$ is calculated.

---

[5] Note: $\log_2 2$ is one bit and the maximum I is $\log_2 s$ bits. Brillouin shows that when the f_i are large, say 100 each or greater, then $\dfrac{I/n}{\log_2 e}$ will come close in value to Shannon's (1948) entropy which is the same as Rényi's entropy of order one (i.e. α approaching 1). Letter e represents the natural base (2.718281828).

3. Brillouin's **information** I is not a divergence measure, therefore it is not the same kind as Rényi's I or Kullback's 2I.

4. $H = \dfrac{E}{n}$ is not the same as R.H. Whittaker's β diversity. Whittaker's β diversity is Rényi's entropy of order one applied to a specific object chosen by him. It is not a different kind of entropy.

Energy-based mapping

Energy-based entropy cloud

"Cloud" is a metaphor for the three dimensional surface in analytical space anchored on the sampling grid by the nH and H coordinates (Table 4, Figure 3). Where do the nH and H values come from? Example:

Values in Table 3, position 1, R1: T=82, n=17
Values in Table 4, Part 1, position 1, R1:
nH = (82+17) ln (82+17) - 82 ln 82 - 17 ln17= 45.401 nats
H = 45.401/17 = 2.671 nats

Table 4. Part 1: Stand level energy-based entropy (nH) and one resonator energy-based entropy (H) parametrised by the T and n values of Table 3. Maps are displayed in Figure 3. Part 2: Graph code for cloud nH. Part 3: Bench and vegetation type code for the sampling grid. See note on position numbering in Table 3.

PART 1	nH				H			
Position	R1	R2	R3	R4*	R1	R2	R3	R4
1	45.401	45.401	40.181	130.984	2.671	2.671	2.679	2.673
2	44.826	39.803	42.975	127.680	2.637	2.654	2.528	2.606
3	50.699	53.201	46.738	150.663	2.535	2.533	2.597	2.554
4	42.975	42.310	48.292	133.766	2.528	2.489	2.683	2.572
5	54.950	53.629	41.139	149.884	2.389	2.554	2.420	2.457
6	44.812	44.826	51.862	141.653	2.490	2.637	2.470	2.530
7	51.750	38.028	43.615	133.495	2.587	2.716	2.566	2.618
8	48.410	47.314	56.636	152.399	2.548	2.490	2.462	2.498
9	48.654	43.615	50.919	143.303	2.433	2.566	2.425	2.471
10	54.593	46.738	44.231	145.656	2.482	2.597	2.602	2.555
11	43.156	45.258	48.622	137.163	2.398	2.514	2.559	2.494
12	45.907	44.826	45.693	136.478	2.550	2.637	2.539	2.575

*Cell values are nH quantities based on marginal T and n in Table 2.

PART 2

Position	Graph colour Figure 3	Graph colour Figure 4

1	Yellow	Yellow	Brick	Green	Green	Green
2	Yellow	Brick	Brick	Yellow	Yellow	Yellow
3	Green	Green	Yellow	Yellow	Yellow	Yellow
4	Brick	Brick	Yellow	Yellow	Yellow	Green
5	Green	Green	Yellow	Brick	Yellow	Brick
6	Yellow	Yellow	Green	Yellow	Yellow	Brick
7	Green	Brick	Brick	Yellow	Green	Yellow
8	Yellow	Yellow	Yellow	Yellow	Yellow	Brick
9	Yellow	Brick	Green	Brick	Yellow	Brick
10	Green	Yellow	Yellow	Yellow	Yellow	Yellow
11	Yellow	Yellow	Yellow	Brick	Yellow	Yellow
12	Yellow	Yellow	Yellow	Yellow	Yellow	Yellow

PART 3

Position		Bench		Vegetation type		
1	B2	B1	B1	Pol	Pol	Pol
2	B2	B1	B1	Pol	Pol	Pol
3	B2	B2	B1	Pol	Mah	Pol
4	B2	B2	B1	Mah	Mah	Pol
5	B2	B2	B1	Mah	Mah	Mah
6	B2	B2	B2	Mah	Mah	Mah
7	B2	B2	B2	Mah	Mah	Mah
8	B2	B2	B2	Gau	Mah	Mah
9	B3	B2	B2	Gau	Mah	Mah
10	B3	B3	B3	Gau	Mah	Mah
11	B3	B3	B3	Gau	Gau	Mah
12	B3	B3	B3	Gau	Gau	Mah

* Legend to vegetation types types: Pol – Polystichm munitum, Mah – Mahonia nervosa, Gaul – Gaultheria shallon. Regarding types, refer to Orlóci (1965).

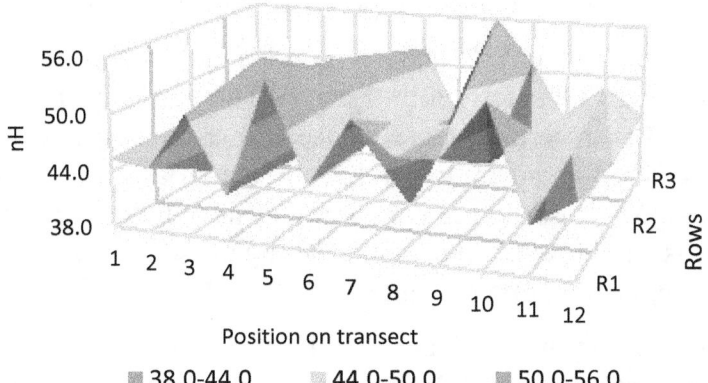

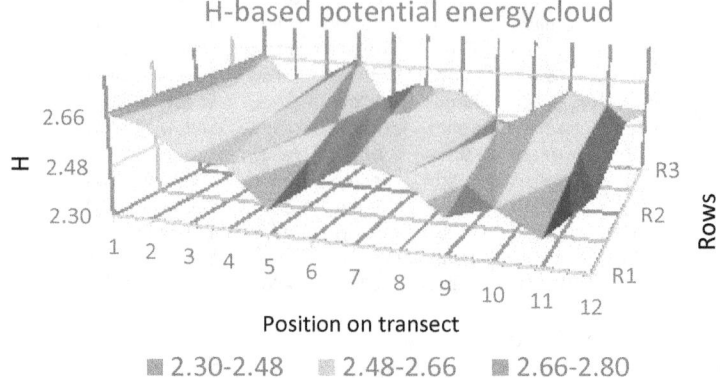

Figure 3. The nH and H cloud. Base: 12x3 sampling grid. Data base: nH and H co-ordinates in Table 4. In all,40 leading taxa and 36 sample plots are used. Position 1: sample plot triplet at north end of the grid (Figure 1). Position 12: triplet at south end of the grid. R1, R2, R3 – parallel rows of 12 sample plots in each.

On first sight of Figure 3, we are struck by the graphs' shape. What does the shape tell us regarding homogeneity of the energy clouds, and beyond, what type of linkage exists to the vegetation types or the height of the benches? The following suggest possible ways to find answers:

1. Parameter correlations
   r(T,nH) =0.629   r(n,nH) =0.974   r(T,H) =0.107   r(n,H) =-0.846

The high correlation of T and n with nH is the direct consequence of nH being parameterised by T and n. The very low correlation of H and T is an indirect consequence of division of nH by n, considering that T and n are logically correlated. Based on the correlation values one may see parsimony in substitutions, n for both nH. This would be at the cost of changing the scale, lowering precision, and making H constant.

2. Homogeneity in the 12x3 nH table
We start with Part 1 of Table 4 included here for quick reference:

Position	R1	R2	R3	Total
1	45.401	45.401	40.181	130.983
2	44.826	39.803	42.975	127.604
3	50.699	53.201	46.738	150.638
4	42.975	42.310	48.292	133.577
5	54.95	53.629	41.139	149.718

6	44.812	44.826	51.862	141.5
7	51.75	38.028	43.615	133.393
8	48.41	47.314	56.636	152.36
9	48.654	43.615	50.919	143.188
10	54.593	46.738	44.231	145.562
11	43.156	45.258	48.622	137.036
12	45.907	44.826	45.693	136.426
Total	576.133	544.949	560.903	1681.985

The shape of the nH cloud suggests an underlying two-dimensional effect whose best fitting plane is not horizontal. We can test the presence of such an effect by showing that a pattern of such type and intensity is unlikely to be the consequence of pure chance. There are different ways of doing the test, but not necessarily with the same outcome:

a. We test the hypothesis that the best fitting plane of the nH cloud in Figure 3 is horizontal. The equation of a best fitting plane is

$$Y = a + bX_1 + cX_2$$

In this Y is the regression estimate of nH, X_1 stand for position 1 to 12, and X_2 represents row 1 to 3. The following are the numerical results of regression analysis:

```
PROGRAM: REGRESSION
============================
FILE X (INDEPENDENT VARIABLES):
x1x2x36.txt
FILE Y (DEPENDENT VARIABLE): yx36.txt
OPTION SELECTED: 2
Number of variables: 2
NUMBER OF OBSERVATIONS: 36
T VALUE AT DF = 33 IS 1.658
INVERSE OF SSCP
   0.00239  -0.00163
  -0.00163   0.04507
ANOVA TABLE
```

X VARIABLE* MEAN VARIANCE
1   6.50000   12.25714
2   1.91667   0.65000
*IF OP=3 THEN X1=X AND X2=X^2

SSCP MATRIX- CORRECTED FOR MEANS
   429.00000   15.50000
    15.50000   22.75000

Source of variation, sum of squares, DF, mean square, F
REGRESSION   14.38309   2   7.19154   0.33597
RESIDUAL   706.37073   33   21.40517
TOTAL   720.75385   35
COEFFICIENT OF DETERMINATION = 0.01996

REGRESSION COEFFICIENTS

	VALUE	STANDARDIZED	T	$P(t_{RND} \geq t)$
B(0)	45.52994			
B(1)	0.18307	0.14124	0.80941	0.2120
B(2)	0.00100	0.00018	0.00102	0.4996

Note:

$t_b = b / \sqrt{0.00239 \times 21.49517}$ and $t_c = c / \sqrt{0.04507 \times 21.40517}$

THE 0.05 PROBABILITY POINT FOR t at DF = 33 IS 1.658
For technical details see Chapter 13 in my "Statistical Ecology" (2012).

We observe that the best fitting plane has ascending slope an arc tan b = 10.37^O North to South direction. The incline from West to East is more severe (0.057^O). These represent the actually observed deviations of the fitted plane from the horizontal position. We have to make a statistical judgment if in fact the deviations are significantly different from zero. Considering the probabilities under $P(t_{RND} \geq t)$ in the regression results, the 0.2120 for b and 04996 for c are more than sufficient to justify the conclusion that the best fitting plane for nH is horizontal. This is the same as saying that whatever real environmental effect is present on the transect, it does not tilt the best fitting nH plane significantly.

b. In second test our hypothesis is that the nH values ground pattern is not predictable from observation of the vegetation types' ground pattern. We approach the problem from an information theoretical point of view as a divergence problem in the context of discriminant contingency table analysis. The contingency table for this is constructed basic on the records in Part 2 of Table 4:

nH 1-12	Pol	Mah	Gaul	Total
Brick	2	5	0	7
Yellow	6	9	6	21
Green	1	6	1	8
Total	9	20	7	36

The test criteria include Rajski's metric (raj), the coherence coefficient (coh), and Kullback's minimum discrimination information statistics (2I). I refer for details and references on these to my "Statistical Ecology" text (2012) for details.

Invoking Rényi's definitions for entropy of order one and information of order one, and the information quantities associated with them, we write $raj = \dfrac{Hj - Hm}{Hj}$ for Rajski's metric, $coh = \sqrt{1 - raj^2}$ for the coherence coefficient and 2I for Kullback's minimum discrimination information statistics. The elementary quantities identity follows their use by Orlóci (2006). The Venn diagram specifies their additivity. For example to get Hm subtract Hj from Hbr + Hbc:

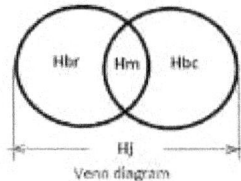

Venn diagram

The following includes the results of information theory based canonical contingency table analysis. The entries in grey are given for interpretation as a student assignement:

α	Unit	Hbr	Hwr	Hbc	Hwc	Hj	Hm	HbrR
1	bits	1.3951	1.3117	1.4304	1.2764	2.7069	0.1187	0.8802
1	nats	0.9670	0.9092	0.9915	0.8847	1.8763	0.0823	0.6101
	HwrR	HBcR	HwcR	HjR	raj	coh	2I*	df
	0.8276	0.0902	0.8053	0.8539				
	0.5736	0.0625	0.5582	0.5919	0.9561	0.2929	5.926	4

*Mminimum discrimination information: $2I = 2 \times T \times Hm = 2 \times 36 \times 0.0823 = 5.926$

Abbreviations: Hbr – entropy of order one between rows, Hwr - within rows, Hbc - between columns; Hwc -, Hj – joint entropy, Hm – mutual entropy, raj – Rajski's metric, coh – Coherence coefficient; 2I – minimum discrimination information statistics; df – degrees of freedom. See for further guidance Orlóci (206, chapters 4, 21 in 2012).

In this test the computational tasks and results are rather complex. We select to interpret the statistics already mentioned:

1. Rajski's metric is almost at its possible maximum and correspondingly the coherence coefficient is low. These indicate weak association between nH and vegetation types.

2. We take 2I as a Chi-squared variate with a single parameter df = 4. The probability of an at least as extreme Ci-squared value as great or greater than 5.929 occurring by pure chance, under the condition of no association, is 0.2049. The verdict is unequivocal: nH and the vegetation types are statistically independent. In other words the ground pattern of vegetation types have no significant predictive value for the energy-based entropy level of the vegetation (in the site of the sampling grid).

c. The set up and analysis in the third test is similar to the second, except for the contingency table which we construct from the records in Part 3 of Table 4:

nH	B1	B2	B3	Total
Brick	3	6	0	9

	Yellow	4	7	9	20
	Green	0	6	1	7
	Total	7	19	10	36

Alpha	Unit	Hbr	Hwr	Hbc	Hwc	Hj	Hm	HbrR
1	bits	1.4304	1.1851	1.4593	1.1562	2.6156	0.2742	0.9025
1	nats	0.9915	0.8214	1.0115	0.8014	1.8130	**0.1901**	0.6256
	HwrR	HBcR	HwcR	HjR	raj	coh	2I	df
	0.7477	0.092	0.7295	0.8251				
	0.5183	0.0638	0.5057	0.5719	0.8951	0.4457	13.6844	4

In this case the coherence coefficient is numerically higher than in the previous case, but Rajski's metric is still very high. Which way should we decide? $2I = 13.6844$ will help in making the statistically best decision. At given 4 degrees of freedom the probability of an at least as large 2I value as the observed 13.6844 occurring by chance alone under the expectation of no association is 0.0084. To this extent, we should consider bench height statistically significant, albeit numerically a rather weak indicator of nH.

d. Vegetation type and bench height are independent. Let us see what is indicated by the 3 x 3 contingency table:

B on Type	Pol	Mah	Gau	Total
B1	6	1	0	7
B2	3	15	1	19
B3	0	4	6	10
Total	9	20	7	36

Alpha	Unit	Hbr	Hwr	Hbc	Hwc	Hj	Hm	HbrR
1	bits	1.4304	0.8955	1.4593	0.8667	2.326	0.5637	0.9025
1	nats	0.9915	0.6207	1.0115	0.6008	1.6123	0.3907	0.6256
	HwrR	HBcR	HwcR	HjR	raj	coh	2I	df
	0.565	0.092	0.5468	0.7338	0.7576	0.6526		
	0.3916	0.0638	0.3790	0.5086	0.7576	0.6526	28.132	4

Clearly, the indicators negate the proposition of independence. We conclude that the vegetation type classification of the sample plots is a reliable indicator of their bench-based classification.

What can we say from the results of this section in the crucial matter of the vegetation types' indicator value for nH? The message is unclear. It is inconsistent with what should be expected based on simple syllogism: bench height is a significant indicator of the nH energy level in the vegetation stand, vegetation type has highly significant association with bench height, yet vegetation type is not a significant indicator of nH? Perhaps the roots of the dilemma reside in the incompatibility of the logic in vegetation classification with the logic of how nH is parameterised. The two logics do not intersect. To see this consider that vegetation classification which produced the types used differential species (see Orlóci 1965). This is in sharp contrast with nH whose parameters are independent from species identities in the sample.

The w_{AB} cloud and its allies

The logical next step, after nH cloud mapping, should be the parameterisation and mapping of w_{AB}. We mentioned already that is a measure instability with minimum 0 at complete stability and maximum 0.5 at complete instability. We note further that $2w_{AB}$ a squared probability which we convert to probability by taking its square root:

$$\omega_{AB} = \sqrt{2w_{AB}}$$

Considering that the low instability level implies high stability level, and that the values rage from 0 to 1, the negative logarithm reverses magnitudes, we call $m\omega_{AB} = -\ln \omega_{AB}$ the *unit stability moment*. Conversely, we call $-m\omega_{AB} = -\ln(1-\omega_{AB})$ the *unit instability moment*. These have to be multiplied by n to obtain *total moments*. The moments have units in nats.

We can use nH or H to calculate values for w_{AB}, or we may start with P values. Taking the first H value from Part 1 in Table 4, we have:

H	2.670662	1-P	0.930794	$m\omega_{AB}$	0.70848437 nat
P	0.069206	$(1-P)^2$	0.866377	$-m\omega_{AB}$	0.67804167 nat
P^2	0.004790	w_{AB}	0.128834		

Table 5 contains the w_{ab} values and the unit moments for 36 vegetation stands. These are mapped in Figure 5.

Table 5. Fisrst block: values of the H-based instability index w_{AB} mapped in Figure 5A onto the 36 nodes of the sampling grid. Second and third block: values of the H-based unit stability moment $m\omega_{ab}$ and unit the instability moment $-m\omega_{ab}$ mapped in Figures 5B and 5C.

Bloc 1: Instability level w_{AB}

Position	R1	R2	R3	Total	Mean
1	0.128834	0.128834	0.127872	0.38554	0.12851333
2	0.132924	0.130892	0.146905	0.410721	0.136907
3	0.145967	0.146173	0.13795	0.43009	0.14336333
4	0.146905	0.152238	0.127382	0.426525	0.142175
5	0.166598	0.143473	0.162034	0.472105	0.15736833
6	0.152136	0.132924	0.154915	0.439975	0.14665833
7	0.139104	0.123494	0.141931	0.404529	0.134843
8	0.144245	0.152044	0.155928	0.452217	0.150739
9	0.16018	0.141931	0.161342	0.463453	0.15448433
10	0.153249	0.13795	0.137281	0.42848	0.14282667
11	0.165342	0.148739	0.142779	0.45686	0.15228667
12	0.143917	0.132924	0.145489	0.42233	0.14077667
Marginal	1.779401	1.671616	1.741808	5.192825	

Block 2. Unit stability moment $m\omega_{AB} = -\ln \omega_{AB}$ where $\omega_{AB} = \sqrt{2w_{AB}}$

Position	R1	R2	R3	Mean
1	0.70848437	0.70848437	0.70463568	0.70720143
2	0.72485356	0.71671947	0.78097875	0.74081001
3	0.77720113	0.77803057	0.74499187	0.76672621
4	0.78097875	0.80250108	0.70267547	0.76195032
5	0.86093011	0.76716711	0.84227316	0.82328767
6	0.80208867	0.72485356	0.81333655	0.77998513
7	0.74962074	0.68712311	0.76097005	0.73253896
8	0.77027156	0.80171672	0.81744292	0.79644343
9	0.83471889	0.76097005	0.83945195	0.81159181
10	0.80659053	0.74499187	0.74230936	0.764569
11	0.85578683	0.78837133	0.76437742	0.80269786
12	0.76895241	0.72485356	0.77527689	0.75633407

Block 3: Unit stability moment $^{-}m\omega_{AB} = -\ln(1-\omega_{AB})$ where $\omega_{AB} = \sqrt{2w_{AB}}$

Position	R1	R2	R3	Mean
1	0.67804167	0.67804167	0.68178917	0.67928772
2	0.66241528	0.67011777	0.61241099	0.64765312
3	0.61561376	0.61490862	0.6438584	0.62461295
4	0.61241099	0.59458151	0.68370883	0.6287747
5	0.54951219	0.62423062	0.56340095	0.57800948
6	0.59491662	0.66241528	0.58586576	0.61325124
7	0.63969312	0.69920776	0.62963354	0.65524848
8	0.62154743	0.59521907	0.58260687	0.59952912
9	0.56915496	0.62963354	0.56554088	0.5872577
10	0.59127202	0.6438584	0.64628909	0.62648816
11	0.55329602	0.6062075	0.62665506	0.59442169
12	0.62268568	0.66241528	0.61725381	0.63371669

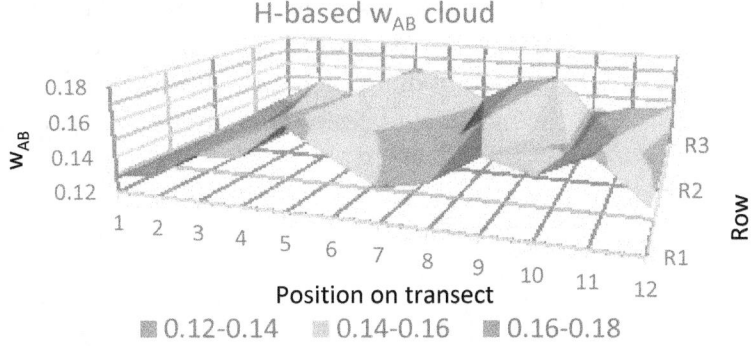

Figure 5A. Energy structural instability level in the 36 stands over the 12x3 sampling grid. The w$_{AB}$ values are taken from Table 5, Block 1.

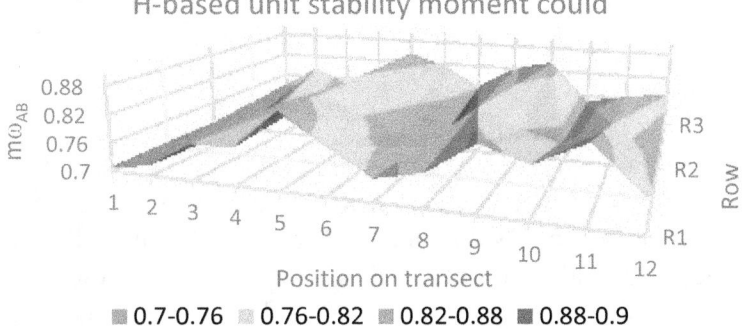

Figure 5B. Energy structural unit stability moment level in the 36 stands over the 12x3 nodes of the sampling grid. The mω$_{ab}$ values are taken from Table 5, Block 2.

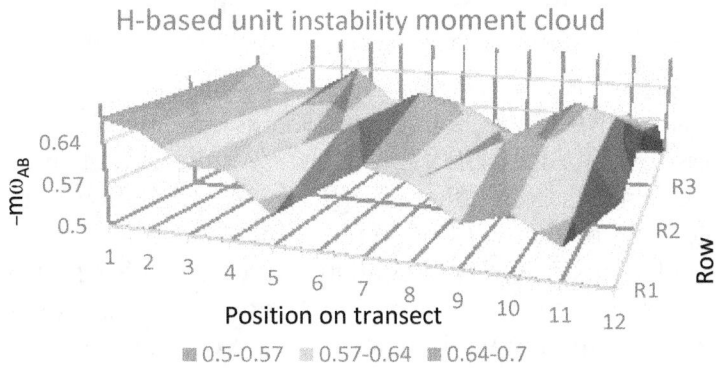

Figure 5C. Energy structural unit instability moment level in the 36 stands over the 12x3 sampling grid. The mω$_{ab}$ values are taken from Table 5, Block 3.

The w_{AB} cloud in Figure 5A has a definitely irregular, periodic shape. It appears to have an ascending component from position 1 to position 12. This component is picked up by the line graph below drawn for the marginal mean w_{AB} of Block 1 in Table 5:

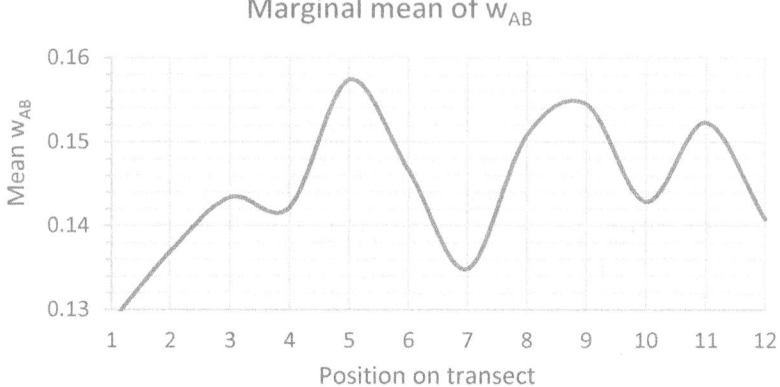

The question is if such a graph could not have originated in a random process with points dispersed by chance around an ascending straight line. A two-step statistical test will tell us if the graph is connecting random points or depicting real periodicity on an ascending path. Two hypotheses define the course of the test:

Hypothesis 1. The slope of the best fitting regression line does not differ significantly from zero.

Hypothesis 2. The residual portion of w_{AB} after the effect of linear regression is removed is randomly distributed about the regression line.

The basis of the test in both case is regression analysis which we perform by TABLECURVE. To interpret the output in entirety the reader needs to find the manual for TABLECURVE. The terminology he or she will find in the manual have overlap with the technical terms in Orlóci's Statistical Ecology (2012). Both resources are downloadable from the Internet. It is very likely that students not already well versed in advance regression analysis will require additional sources of instruction..

Regarding Hypothesis 1, we present the graphs and complete numerics of regression analysis:

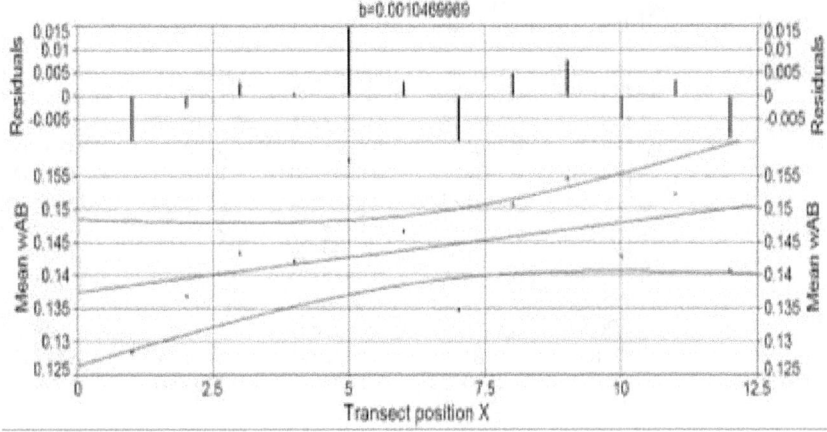

Rank 2 Eqn 8162 [Line Robust Medium, Lorentzian Errors] y=a+bx

r² Coef Det	DF Adj r²	Fit Std Err	F-value
0.1962453894	0.0176332537	0.0080115453	2.4416082567

Parm	Value	Std Error	t-value	95% Confidence Limits		P>|t|
a	0.137440211	0.004930765	27.87401200	0.126453781	0.148426641	0.00000
b	0.001046892	0.000669959	1.562620652	-0.00044587	0.002539654	0.14921

Area Xmin-Xmax	Area Precision		
1.5866950890	0.0000000000		
Function min	X-Value	Function max	X-Value
0.1384871043	1.0000015044	0.1500029135	12.000000000
1st Deriv min	X-Value	1st Deriv max	X-Value
0.0010468919	3.8849607953	0.0010468919	1.9459082306
2nd Deriv min	X-Value	2nd Deriv max	X-Value
-2.80011e-11	5.8213309908	2.800109e-11	4.1467477622

Procedure	Minimization	Iterations		
LevMarqdt	Least Squares	7		
r² Coef Det	DF Adj r²	Fit Std Err		
0.1962453894	0.0176332537	0.0080115453		
Source	Sum of Squares	DF	Mean Square	F Statistic	P>F
Regr	0.00015671428	1	0.00015671428	2.44161	0.14922
Error	0.00064184858	10	6.4184858e-05		
Total	0.00079856286	11			

The probability under P>|t| in the table of parameters reads 0.14921. This is telling us that the angle of ascent is 0.06^O, directly obtained as the arctangent of the value of the regression coefficient b, is not sufficiently large to induce us to negate Hypothesis 1.

We also note that 6 of the 12 points is located outside or at the 95% confidence limits. These encourage us to perform the test on Hypothesis 2, but at this time residuals of w_{AB} are used. The residuals are in the fourth of the table below:

Position X	Mean w_{AB}	M w_{AB} pred.	Residuals	95% Conf. lim. pred.	
1	0.1285133	0.1384862	-0.009973	0.1287928	0.1481796
2	0.136907	0.1395332	-0.002626	0.1310669	0.1479995
3	0.1433633	0.1405802	0.0027832	0.1332418	0.1479185
4	0.142175	0.1416272	0.0005478	0.1352647	0.1479897
5	0.1573683	0.1426742	0.0146942	0.1370556	0.1482927
6	0.1466583	0.1437212	0.0029372	0.1385143	0.148928
7	0.134843	0.1447682	-0.009925	0.1395613	0.149975
8	0.150739	0.1458151	0.0049239	0.1401966	0.1514337
9	0.1544843	0.1468621	0.0076222	0.1404996	0.1532246
10	0.1428267	0.1479091	-0.005082	0.1405708	0.1552475
11	0.1522867	0.1489561	0.0033305	0.1404898	0.1574224
12	0.1407767	0.1500031	-0.009226	0.1403097	0.1596965

With the aim of capturing periodic change, we fit a Fourier series polynomial to the residuals. The regression graph and numerics are below:

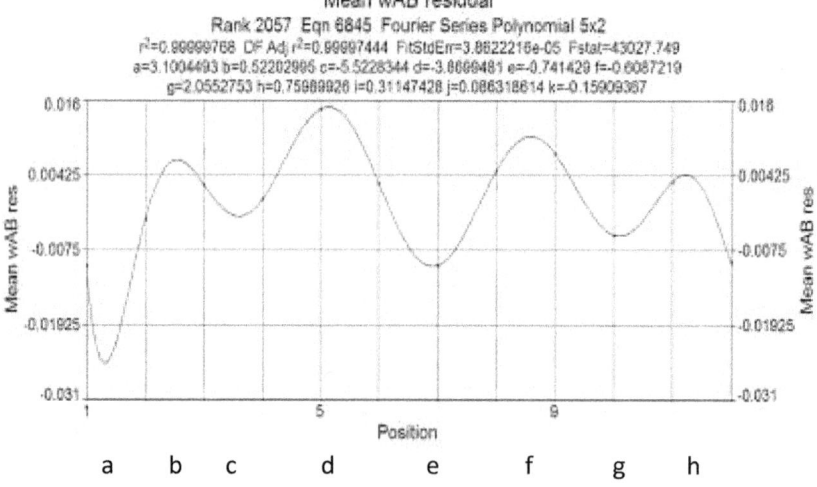

Mean wAB residual

Rank 2057 Eqn 6845 Fourier Series Polynomial 5x2
r^2=0.99999768 DF Adj r^2=0.99997444 FitStdErr=3.8622216e-05 Fstat=43027.749
a=3.1004493 b=0.52202995 c=-5.5228344 d=-3.8699481 e=-0.741429 f=-0.6087219
g=2.0552753 h=0.75989926 i=0.31147428 j=0.086318614 k=-0.15909367

Rank 2057 Eqn 6845 Fourier Series Polynomial 5x2

r^2 Coef Det	DF Adj r^2	Fit Std Err	F-value
0.9999976759	0.9999744352	3.862222e-05	43027.749034

Parm	Value	Std Error	t-value	95% Confidence Limits		P>|t|
a	3.100449316	0.024358247	127.2854026	2.790948438	3.409950194	0.00006
b	0.522029945	0.001709195	305.4244683	0.500312565	0.543747325	0.00001
c	-5.52283445	0.043185095	-127.887514	-6.07155311	-4.97411579	0.00006
d	-3.86994807	0.029755544	-130.058050	-4.24802811	-3.49186803	0.00006
e	-0.74142900	0.002424251	-305.838346	-0.77223203	-0.71062597	0.00001

f	-0.60872190	0.001974293	-308.323988	-0.63380767	-0.58363813	0.00001
g	2.055275335	0.015337975	133.9991362	1.860387891	2.250162780	0.00006
h	0.759899257	0.005402486	140.6573390	0.691254168	0.828544346	0.00005
i	0.311474283	0.001007668	309.1041299	0.298670649	0.324277916	0.00001
j	0.086318614	0.000270821	318.7291618	0.082877504	0.089759723	0.00001
k	-0.15909367	0.001010909	-157.376828	-0.17193849	-0.14624885	0.00004

Area Xmin-Xmax	Area Precision		
0.0042926793	1.554874e-14		
Function min	X-Value	Function max	X-Value
-0.025204764	1.3149196204	0.0150070269	5.1265321187
1st Deriv min	X-Value	1st Deriv max	X-Value
-0.022144032	6.0187298604	0.0422387227	1.7514646646
2nd Deriv min	X-Value	2nd Deriv max	X-Value
-0.054331409	11.412360792	0.4991806929	1.0000015044

Soln Vector	Covar Matrix	
GaussElim	LUDecomp	
r^2 Coef Det	DF Adj r^2	Fit Std Err
0.9999976759	0.9999744352	3.862222e-05

Source	Sum of Squares	DF	Mean Square	F Statistic	P>F
Regr	0.00064183443	10	6.4183443e-05	86055.5	0.00001
Error	1.4916756e-09	2	7.458378e-10		
Total	0.00064183592	12			

The polynomials' extreme points are mapped in the following table in relation to dominant vegetation type and bench height in the positional triplets (Table 4):

	w_{AB} minima				w_{AB} maxima			
	a	c	e	g	i	b	d	f	h
Vegetation type	Pol	Mah	Mah	Mah	Gaul	Pol	Mah	Mah	Gaul
Bench	B1	B2	B2	B3	B3	B2	B2	B2	B3

The regression statistics indicate a perfect fit of the Fourier polynomial ($r^2 = 0.9999$). All regression coefficients differ significantly from zero (see entries under P>|t| in the parameters table). In other words none of the 95% confidence limit about the regression coefficients capture zero. A word of warning is in order before we ask what to make of the message read from regression analysis.

It is best to regard the regression curve as an artefact, a direct consequence of selecting a Fourier series polynomial, until we can show that the extreme points are representing something inherently meaningful in ecology. The truth is that we do not have any record on the basis of which we could close in on the forcing ecological factors. What makes us to direct our effort in another direction is the impossibility of a second survey in the site which we would be considering consistent with normal ecological practice called *successive approximation*. Yet, the situation in which we find ourselves is not completely hopeless. There are pointers that we can go after, identify and interpret.

The correlation values of w_{AB} and selected parameters lead us to interesting fact:

Parameter	T	n	nH	H	P
r(Par, W_{AB})	- 0.234	**0.638**	0.417	-0.999	1.000

We recall that w_{AB} derived from the unit resonator's potential energy H which is the mean value given by En^{-1}. The functional negative correlation of w_{AB} with H and the high positive correlation with n suggest that energy instability increases in the complex with the number of resonators. This can be rephrased: the vegetation stand's energy structural instability is simply proportional to the number of the community elements (species) and inversely proportional to the one-element (species) energy-based entropy level. Taking this a step further, we can conclude that a species rich stand is more prone to flipping into one of its ghost states by pure chance than a species poor stand. Is this surprising? It is. Does it make sense? It does. But we have to think of the vegetation stand's energy structural state as one in a continuum through time. The chronological process targets its own attractor around which the ghost states cluster in a normal manner.

What are the mechanism in nature selecting from the available ghost states? I mention only one which is intrinsic and effective and do not depend on catastrophic events. This mechanism was first identified by the father of ecology, the Austro-Hungarian Kerner von Marilaun (1864). Modern ecology's term for the process is *facilitation.*

Further on ghost states

Superposition

We already discussed the idea of superposition and used P to denote the observed potential energy structural state's probability of occurring by chance. The complement 1-P is the probability that the energy structure flips by chance into a ghost state. Only values P<1 are attainable in Nature. We further note that since any energy structural state is inherently unstable (P<1), any complex can flip into another energy state by pure chance at the rate of 1-P. We call this *superposition.*

Emergent potential energy

We define the emergent potential energy for any vegetation complex A in the manner of

$$gnH(A) = dnH(A) - nH(A)$$

such that

$$dnH(A) = nH(A + B) - nH(B)$$

We assume that A, a complex or a sere, can be logically concatenated with B, another complex or sere. The linkage in the present case is provided by spatial catenation on the transect.

The terms in the two eqation have been defined. The example

below illustrates the compotation on data taken from Table 6:

Plot 1, R1	T*	82
	n*	17
	nH*	45.40126000
	H*	2.67066236
40 x 36 block	T	2784
	n	661
	nH	1684.35154311
Grand T minus Plot 1 T	dT	2702
Grand n minus Plot 1 n	dn	644
	nH	1638.81309500
	dnH	45.53844810
Emergent gnH		0.13718810

*Values from Table 3, 4

The dnH and gnH clouds are presented in Figures 6 and 7.

Table 6. Data sets for the calculation of the emergent potential energy cloud gnH and the data for the T and n values are reproduced for the sake of convenience from Table 3 and the Table 4.

Position	T matrix			n matrix		
	R1	R2	R3	R1	R2	R3
1	82	82	73	17	17	15
2	79	71	70	17	15	17
3	83	87	80	20	21	18
4	70	67	88	17	17	18
5	81	89	62	23	21	17
6	71	79	81	18	17	21
7	88	71	73	20	14	17
8	80	75	88	19	19	23
9	74	73	77	20	17	21
10	86	80	76	22	18	17
11	64	73	81	18	18	19
12	76	79	75	18	17	18

Grand totals: T=2784, n=661

Pos	dT			dn		
	R1	R2	R3	R1	R2	R3
1	2702	2702	2711	644	644	646
2	2705	2713	2714	644	646	644
3	2701	2697	2704	641	640	643
4	2714	2717	2696	644	644	643
5	2703	2695	2722	638	640	644
6	2713	2705	2703	643	644	640
7	2696	2713	2711	641	647	644
8	2704	2709	2696	642	642	638
9	2710	2711	2707	641	644	640

10	2698	2704	2708	639	643	644
11	2720	2711	2703	643	643	642
12	2708	2705	2709	643	644	643

nH(A)

Pos	R1	R2	R3
1	1638.813	1638.813	1644.033
2	1639.454	1644.46	1641.373
3	1633.651	1631.148	1637.591
4	1641.373	1642.012	1635.883
5	1629.115	1630.722	1643.074
6	1639.509	1639.454	1632.424
7	1632.585	1646.108	1640.734
8	1635.941	1637.005	1627.63
9	1635.565	1640.734	1633.274
10	1629.709	1637.591	1640.094
11	1640.996	1639.083	1635.728
12	1638.444	1639.454	1638.657

dnH cloud

Pos	R1	R2	R3
1	45.53845	45.53845	40.31886
2	44.89745	39.89154	42.97828
3	50.70048	53.20357	46.76009
4	42.97828	42.33983	48.46903
5	55.23609	53.62957	41.27716
6	44.84297	44.89745	51.92725
7	51.7661	38.24356	43.61737
8	48.41039	47.34611	56.72118
9	48.78682	43.61737	51.07779
10	54.64296	46.76009	44.25709
11	43.35584	45.2685	48.62346
12	45.90733	44.89745	45.69432

gnH cloud

Pos	R1	R2	R3
1	0.137188	0.137188	0.137509
2	0.071029	0.088695	0.00357
3	0.00181	0.002371	0.0221
4	0.00357	0.030256	0.176814
5	0.286187	0.00034	0.137742
6	0.031298	0.071029	0.065377
7	0.016218	0.215281	0.002663
8	6.17E-07	0.032353	0.085336
9	0.132576	0.002663	0.158968
10	0.049589	0.0221	0.025696
11	0.200067	0.010519	0.001167
12	4.56E-05	0.071029	0.000864

$$P = 1 - e^{-gnH}$$

Pos	R1	R2	R3
1	0.12819369	0.12819369	0.12847349

2	0.06856512	0.08487536	0.00356364
3	0.00180836	0.00236819	0.02185758
4	0.00356364	0.02980287	0.16206437
5	0.24887786	0.00033994	0.12867654
6	0.03081329	0.06856512	0.06328574
7	0.0160872	0.19368517	0.00265946
8	6.2E-07	0.03183524	0.08179628
9	0.12416363	0.00265946	0.14697634
10	0.04837954	0.02185758	0.02536867
11	0.1813241	0.01046387	0.00116632
12	4.56E-05	0.06856512	0.00086363

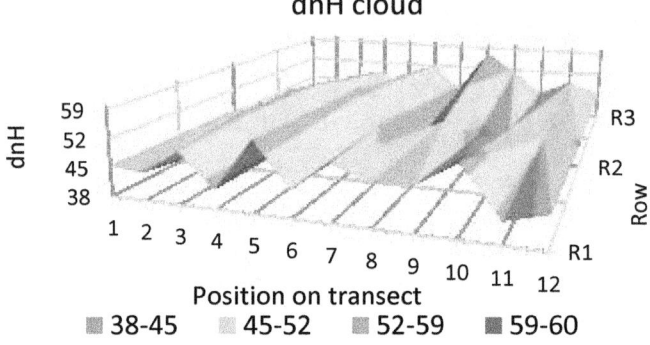

Figure 6. The dnH cloud for 36 sample plots of the 12x3 sampling grid. Basic data in Table 6.

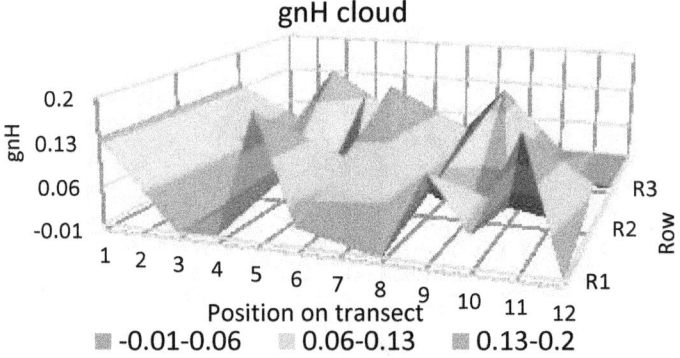

Figure 7. The gnH cloud for 36 sample plots of the 12x3 sampling. Basic data in Table 6.

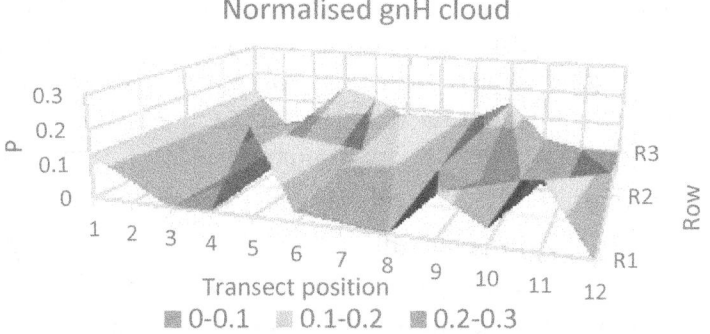

Figure 7. The normalised gnH cloud for 36 sample plots. The value of $P = 1 - e^{-gnH}$ ranges from 0 to 1. P discriminates against the proposition that gnH has zero expectation. See basic data in Table 6.

The dnH cloud is similar to the nH cloud. The difference has magnitude exactly the size of the gnH cloud. The last block in Table 6 contains the normalised gnH values. Normalisation is by negative exponentiation. We take the one-complement of this to have a direct measure of the strength by magnitude of the basic proposition that gnH has zero expectation.

Looking at Table 6 the gnH values may appear excessively small when compared to the nH values. This is a fact, but it is important to us to know whether the small gnH values represent significant deviations from zero. If they do then the presence of *emergent potential energy* is established. We can test for zero expectation statistically. The test can be done in several ways but not necessarily leading to he same conclusion:

1. Each $P = 1 - e^{-gnH}$ value in the last block of Table 6 is specific to the gnH value in the same position of its block. Note that as gnH increases so does P and the proposition of gnH not being significantly different from zero becomes less tenabl. An example will clarify this. Let P=0.1 be our threshold value for significance, implying that any P at least 0.1 is declared significant at 1-P Type I error. At this threshold, 10 out of the 36 gnH values should be regarded as significantly different from zero.

b. The question of statistical significance may be posed in yet another way. We ask the same question, namely, what is the probability that a gnH value at least as extreme as the observed can occur by pure chance under, but we get the answer by way of D.W. Goodall's reasoning in connection of his "probabilistic index" (Orlóci 2012). The relevant test criterion is $\chi^2 = 2 \ln v$ in which v is the number of gnH values in the sample greater or equal to the target gnH. For example, we find gnH=0.286187, the 5th element under R1 in the second last block of Table 6. For this $v = 36$ and $\chi^2 = 7.1670379$. The chi-squared has an associated probability of 0.0278 to be matched or exceeded by a random value of chi-squared at 2 degrees of freedom. We declare gnH=0.286187 significant.

Table 7. Chi-squared numerics. The analysis is performed on the gnH matrix in Table 6. The smaller the value in the P column, the more unique is the observed value of gnH.

R1	gnH	v	Chi-squared	P, df=2
1	0.137188	29	6.73459166	0.03448276
2	0.071029	24	6.35610766	0.04166667
3	0.00181	6	3.58351894	0.16666667
4	0.00357	11	4.79579055	0.09090909
5	0.286187	36	7.16703788	0.02777778
6	0.031298	18	5.78074352	0.05555556
7	0.016218	13	5.12989871	0.07692308
8	6.17E-07	1	0	1
9	0.132576	27	6.59167373	0.03703704
10	0.049589	20	5.99146455	0.05
11	0.200067	34	7.05272105	0.02941176
12	0.0000456	2	1.38629436	0.5
R2				
1	0.137188	28	6.66440902	0.03571429
2	0.088695	26	6.51619308	0.03846154
3	0.002371	7	3.8918203	0.14285714
4	0.030256	17	5.66642669	0.05882353
5	0.00034	3	2.19722458	0.33333333
6	0.071029	23	6.27098843	0.04347826
7	0.215281	35	7.11069612	0.02857143
8	0.032353	19	5.88887796	0.05263158
9	0.002663	9	4.39444915	0.11111111
10	0.0221	15	5.4161004	0.06666667
11	0.010519	12	4.9698133	0.08333333
12	0.071029	22	6.18208491	0.04545455

R3

1	0.137509	30	6.80239476	0.03333333
2	0.00357	10	4.60517019	0.1
3	0.0221	14	5.27811466	0.07142857
4	0.176814	33	6.99301512	0.03030303
5	0.137742	31	6.86797441	0.03225806
6	0.065377	21	6.08904488	0.04761905
7	0.002663	8	4.15888308	0.125
8	0.085336	25	6.43775165	0.04
9	0.158968	32	6.93147181	0.03125
10	0.025696	16	5.54517744	0.0625
11	0.001167	5	3.21887582	0.2
12	0.000864	4	2.77258872	0.25

Further on gnH

An interesting piece of fact is brought to light by the correlation values below:

Entity	r(gnH, Entity)
T	-0.115
n	-0.007
nH	-0.058
H	-0.055
P	0.102
w_{AB}	0.098

Clearly, gnH is linearly independent from the energy structure's defining parameters. What does this tell us? Emergent energy is not predictable.

Phylogeny's local footprint

After considerations of the spatioenvironmental component (E_{Env}), we now turn to the second potential energy component (E_{Phy}) issuing from energy spent in the evolutionary process. The basic model of the analysis is the hierarchical relevé (see Orlóci 2013a,c). The basis of such a relevé is a map of the taxonomical structure of the vegetation stand in the manner as seen in Table 2. We take the taxonomic structure as modern plant systematics hands it to us without arbitrary changes, and use it as local proxy for the phylogenetic tree.

The taxonomic code distinguishes hierarchical levels, including L1 (7 Classes), L2 (16 Orders), L3 (21 Families), and L4 (35 Genera). The proxy evolutionary tree thus includes 7 nodes on level L1, 16 nodes on level L2, 21 nodes on level L3, and 35 nodes on level L4. The baseline level L0 has 40 nodes with a single branch to each taxon. The classification as given in Table 1 reflects the state of plant systematics as it filtered down into the field manuals.

In the hierarchical relevé there is a value attach to each node, called *cumulant*. This is the sum of species quantities accumulated up through the hierarchical pathway to the node. The cumulants use data from Table 8. Note that in this table the marginal sum of any species is increased by 1 on any floodplain bench for which the original value is zero. The one for zero substitutions assume that all species are present on all benches, only some were missed by chance in the course of sampling. The substitution facilitates the analysis without having much effect on the final results.

Table 8. Modified bench totals for 45 sample plots (as in Table 2). Find explanation in the text.

#	Species	B1	B2	B3	B4
1	Thuja plicata (shrub)	1	1	8	10
2	Holodiscus discolor	1	1	13	15
3	Mnium spinulosum	15	2	1	18
4	Lonicera ciliata	1	1	21	23
5	Claytonia sibirica	1	5	17	23
6	Rosa gymnocarpa	1	1	23	25
7	Spiraea douglasii	2	8	15	25
8	Amelanchier florida	26	1	1	28
9	Vaccinium parviflorum	3	8	16	27
10	Acer macrophyllum	30	1	1	32
11	Galium triflorum	23	3	5	31
12	Acer macrophyllum	26	6	1	33
13	Osmorhiza chilensis	1	5	30	36
14	Circaea alpina	1	2	33	36
15	Pteridium aquilinum	25	10	1	36
16	Vaccinium membranaceum	18	18	1	37
17	Smilacina stellata	1	4	34	39
18	Tsuga heterophylla	32	14	1	47
19	Linnaea borealis	47	5	1	53
20	Chimaphila umbellata	1	39	15	55
21	Goodyera oblongifolia	1	54	1	56
22	Dicentra formosa	1	61	1	63
23	Trientalis latifolia	7	38	19	64
24	Rubus spectabilis	8	43	16	67
25	Clintonia uniflora	1	82	3	86
26	Mnium insigne	69	17	2	88
27	Symphoricarpos albus	70	18	4	92
28	Lactuca canadensis	25	47	34	106
29	Rhytidiadelphus loreus	55	54	1	110
30	Disporum hookerii	69	50	1	120
31	Gaultheria shallon	1	43	82	126
32	Polystichum munitum	4	65	72	141
33	Pachistima myrsinites	105	47	1	153
34	Achlys triphylla	63	94	17	174
35	Eurhynchium oreganum	42	99	51	192
36	Mahonia nervosa	37	132	43	212
37	Thuja plicata	38	122	52	212
38	Hylocomium splendens	23	123	73	219
39	Acer macrophyllum (shrub)	106	105	29	240
40	Pseudotsuga menziesii	113	121	88	322
	Sum	1094	1550	828	3472

The results are summarized in Table 9.

Table 9a. T, n, nH, and H matrices for hierarchical level L1 (Class) on benches B1, B2, B3. B4 is the pooled sample.

T, L1

Node	B1	B2	B3	B4
1	189	144	127	460
2	204	295	128	627
3	275	240	133	648
4	141	348	179	668
5	72	190	39	301
6	184	258	149	591
7	29	75	73	177

n, L1

Node	B1	B2	B3	B4
1	9	9	9	9
2	5	5	5	5
3	6	6	6	6
4	10	10	10	10
5	4	4	4	4
6	4	4	4	4
7	2	2	2	2

nH, L1

Nodes	B1	B2	B3	B4
1	36.61167	34.22887	33.13428	44.493492
2	23.60419	25.42982	21.30937	29.1774266
3	29.01505	28.20766	24.72489	34.1204798
4	36.80826	45.6385	39.12228	52.0915104
5	15.6706	19.48473	13.30752	21.3097247
6	19.35773	20.69751	18.52382	23.9955926
7	7.41573	9.275115	9.221775	10.9772622

Level L1	B1	B2	B3	B4
n	7	7	7	7
T	1094	1550	828	3472
nH	42.38415	44.81648	40.44123	50.4530832

Table 9 is set up to show the steps in the analysis and the energy parameters used. In brief, we take the T and n matrices of a bench B, compute an nH value for each node of the dendrogram, and create graphs. We use the nH matrix to calculate the specific contributions of each phylogenetic level to the total potential energy footprint of the vegetation stands.

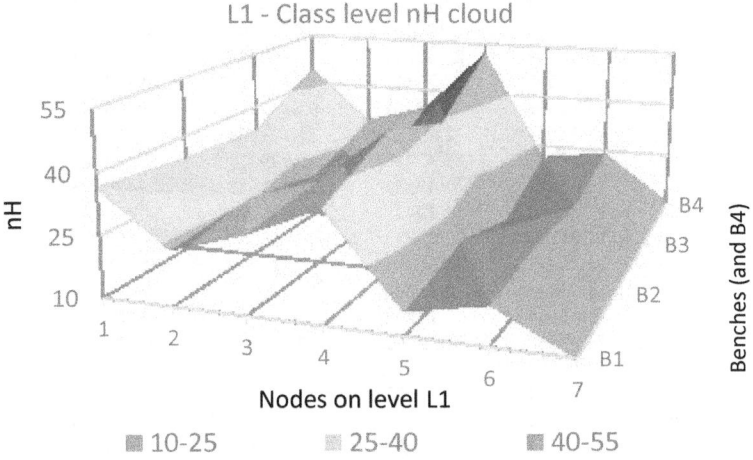

Figure 8a. Potential energy state on the Class level (L1, 7 nodes) of the proxy phylogenetic tree taken from Table 9a. B1, B2, B3 – benches on the floodplain starting at the river. B4 –marginal total. Node identities in random order: 1 - Asterid, 2 - Bryopsida, 3 - Eudicot, 4 - Magnoliopsida, 5 - Monocot, 6 - Pteridopsida, 7 - Pinopsida. Find further information in the text and the external literature.

The graph in Figure 8a identifies the nH parameter related potential energy state of the Class level for each of the 36 vegetation stands. The Asterids and the Magnoliopsida are high contributors. The Pinopsida is lowest. This pattern carries over to all benches and into the pooled sample (B4). The cloud pattern suggests a homogeneous potential energy level on the Class level within the sample plot triplets, and periodic irregularly over the sampling grid through the bench levels.

Further results complete this section, including the nH matrices and graphs for levels L2, L3, and L4 of the proxy phylogenetic tree. Interpretations are left for the reader.

Table 9b. The nH matrix on Class level (L2, 16 nodes) of the proxy phylogenetic tree.

nH,L2 Nodes	B1	B2	B3	B4
1	1.386294	2.703367	4.4176819	4.597281
2	2.772589	8.768685	7.7804863	9.7423659
3	4.238614	4.860711	4.5409243	5.6681413
4	1.386294	2.703367	3.8620648	4.156925
5	5.658707	4.860711	1.3862944	6.0336988
6	7.41573	9.275115	9.221775	10.977262
7	14.05403	9.418467	9.6452322	15.102683
8	14.35407	21.9555	21.497385	25.659753
9	23.60419	25.42982	21.309371	29.177427
10	4.156925	2.249341	2.7033673	4.4499456
11	9.138999	10.39439	3.819085	11.279135
12	1.386294	1.909543	4.5115083	4.597281
13	19.35773	20.69751	18.523825	23.995593
14	13.59364	16.69823	12.109375	18.035232
15	15.45614	17.12238	18.229826	22.406003
16	14.99456	13.89948	10.146818	16.879804
Level L2	B1	B2	B3	B4
n	16	16	16	16
T	1094	1550	828	3472
nH	83.71655	89.25704	79.296394	102.11517

L2 - Order level nH cloud

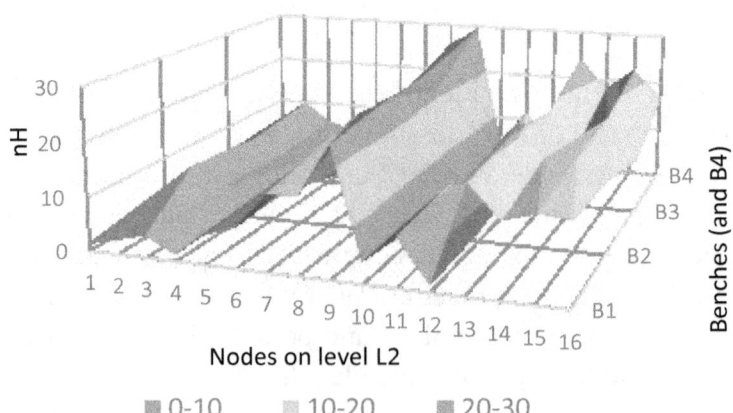

nH

30
20
10
0

Nodes on level L2
1 2 3 4 5 6 7 8 9 10 11 12 13 14 15 16

B4
B3
B2
B1

Benches (and B4)

■ 0-10   ■ 10-20   ■ 20-30

Figure 8b. Potential energy state on the Order level (L2, 16 nodes) of the proxy
phylogenetic tree. Node identities are in random order as in Table 1.

Table 9c. The nH matrix on the Family level (L3, 21 nodes) of the proxy phylogenetic tree. See caption to Table 9a and Figure 8a.

nH,L3 nodes	B1	B2	B3	B4
1	14.995	13.89948	10.14682	16.8798
2	4.1569	2.249341	2.703367	4.449946
3	15.456	17.12238	18.22983	22.406
4	1.3863	4.676274	3.740667	5.016369
5	1.3863	2.703367	3.862065	4.156925
6	10.581	10.439	9.613284	12.44071
7	2.7726	8.699706	7.72413	9.678866
8	3.0142	4.65063	3.970305	5.166655
9	9.499	6.604336	3.365058	9.959334
10	1.3863	5.412792	2.249341	5.460139
11	14.104	16.58161	14.22482	18.48003
12	1.3863	5.119026	1.386294	5.15103
13	9.1731	12.47077	13.53452	15.4688
14	7.4157	9.275115	9.221775	10.97726
15	7.9913	10.25425	8.835364	11.42804
16	5.2413	4.921957	1.386294	5.791647
17	5.6587	4.860711	1.386294	6.033699
18	14.054	9.418467	9.645232	15.10268
19	9.8439	11.4636	8.835364	12.53055
20	4.2386	4.860711	4.540924	5.668141
21	2.7726	5.215532	8.962401	9.275115

Level L3	B1	B2	B3	B4
n	21	21	21	21
T	1094	1550	828	3472
nH	104.21	111.4729	98.42839	128.3306

L3 - Family level nH cloud

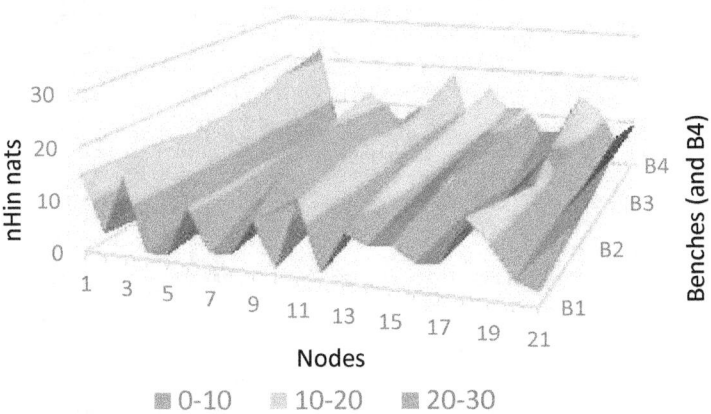

Figure 8c. Potential energy state on the Family level (L3, 21 nodes) of the proxy phylogenetic tree. Node identities are in random order as in Table 1.

Table 9d. The nH matrix on the Genus level (L4, 35 nodes) of the proxy phylogenetic tree. See caption to Table 9a and Figure 8a.

nH, L4 nodes	B1	B2	B3	B4
1	14.99456	13.89948	10.14682	16.8798
2	5.1510296	5.548595	3.862065	6.161923
3	4.2770854	1.386294	1.386294	4.349853
4	1.3862944	4.676274	3.740667	5.016369
5	1.3862944	1.909543	4.511508	4.597281
6	1.3862944	2.703367	3.862065	4.156925
7	1.3862944	5.412792	2.249341	5.460139
8	1.3862944	5.119026	1.386294	5.15103
9	5.2413181	4.921957	1.386294	5.791647
10	4.749481	5.600153	4.941566	6.260095
11	4.156925	2.249341	2.703367	4.449946
12	2.7725887	8.226866	9.742366	10.52534
13	1.3862944	4.998187	1.386294	5.034228
14	4.4812006	3.67395	1.386294	4.860711
15	4.156925	5.816238	5.297278	6.391351
16	4.2386144	4.860711	4.540924	5.668141
17	4.8607112	2.703367	1.386294	4.979667
18	1.3862944	1.386294	4.067963	4.156925
19	4.6243113	5.88658	4.772739	6.358941
20	9.498962	6.604336	3.365058	9.959334
21	1.3862944	2.703367	4.417682	4.597281
23	5.6587072	4.860711	1.386294	6.033699
24	2.5020121	5.18204	5.283579	5.952298
25	5.7317996	5.799911	5.482997	6.776103
26	4.2386144	3.350997	1.386294	4.597281
27	5.0163695	4.998187	1.386294	5.705012
28	1.3862944	1.386294	4.156925	4.238614
29	3.1394889	4.772739	3.803207	5.212118
30	1.3862944	2.502012	4.540924	4.676274
31	1.9095425	3.139489	3.740667	4.238614
32	5.2556043	3.917649	2.502012	5.527204
33	7.9912562	10.25425	8.835364	11.42804
34	3.0141613	4.65063	3.970305	5.166655
35	6.7951014	7.204922	6.393419	8.962401
Level L4	B1	B2	B3	B4
n	34	34	34	34
T	1094	1550	828	3472
nH	152.54495	164.2383	143.2389	191.4542

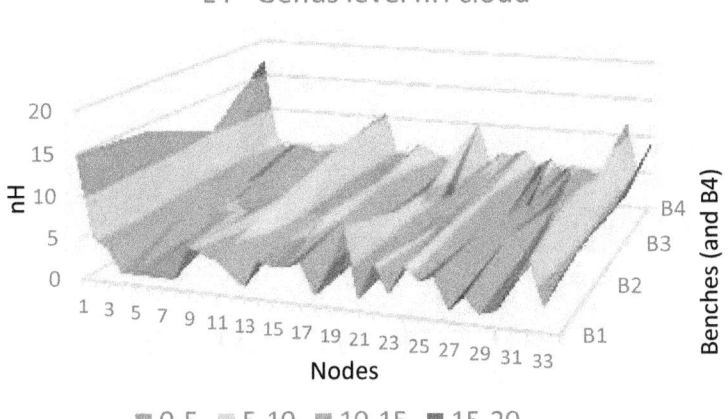

Figure 8d. Potential energy state on the Genus level (L2, 35 nodes) of the proxy phylogenetic tree. Node identities are in random order as in Table 1.

The graphs help to identify high potential energy contributor taxa:

L1 – Asterids, Magnoliopsida, Pinopsida
L2 – Asterales, Eubryales, Pinales
L3 – Sapindaceae, Rosaceae, Pinaceae, Hypnaceae, Caprifoliaceae
L4 – Acer, Holodiscus, Trientalis

Energy structure

Hierarchical levels

Consider the nH totals by floodplain bench and hierarchical level copied from Table 9:

L1 (Class)	B1	B2	B3	B4
n	7	7	7	7
T	1094	1550	828	3472
nH	42.38415	44.81648	40.44123	50.45308
L2 (Order)				
n	16	16	16	16
T	1094	1550	828	3472
nH	83.71655	89.25704	79.29639	102.1152
L3 (Family)				
n	21	21	21	21
T	1094	1550	828	3472
nH	104.2148	111.4729	98.42839	128.3306
L4 (genus)				
n	34	34	34	34
T	1094	1550	828	3472
nH	152.545	164.2383	143.2389	191.4542
L0 (species)				
n	27	34	28	40
T	1094	1550	828	3472
nH	127.278	164.2383	123.2988	218.7738

We summarise nH by floodplain bench x hierarchical level and perform the specified arithmetic:

	Nominal values					Level specific values		
	B1	B2	B3	B4	n	dn	dnH	H
L1	42.3841	44.8165	40.4412	50.4531	7	7	50.4531	7.2076
L2	83.7166	89.2570	79.2964	102.1152	16	9	51.6621	5.7402
L3	104.2148	111.4729	98.4284	128.3306	21	5	26.2154	5.2431
L4	152.5450	164.2383	143.2389	191.4542	34	13	63.1236	4.8557
L0	127.2780	164.2383	123.2988	218.7738	40	6	27.3196	4.5533

The potential energy quantities specific to the different hierarchical levels are given under the column headings dnH and H for the pooled sample (B4). The more ambitious student may calculate the specifics for the individual benches and compare.

Statistically speaking, the specifics are residuals revealed by hierarchical decomposition of the total nH into independent quantities specific to the levels. To summarise:

1. The B columns are benches and the L rows are hierarchical levels. Each floodplain bench is described by its total nH value and specific dnH.

2. The nH value increases with hierarchical level from top (L1) to bottom (L0). This is expected, considering that proliferation of taxa occurs, which shows up as an increase (dnH) in the potential energy level. The specific dnH value lacks monotonicity.

3. Potential energy is cumulative. In other words the dnH values are additive. Therefore we say that a given hierarchical level's dnH is its specific contribution to the total potential energy footprint.

4. Moving from Class to Order (L1 to L2) the dnH value is a huge quantity, less on moving to the next level (L3), and maximal moving from Family to Genera (L3 to L4). The last branching (L4 to L0) adds about the same amount to the total potential energy to the footprint of the complex as the brunching from L2 to L3.

5. Generalization that the total energy footprint of the vegetation complex increases through the levels of the proxy phylogenetic

tree opens up the possibility to incorporate resource-based parsimony into the survey design.

6. The nH (or dnH) quantities are not comparable between different cases. When such comparison is intended, we reach for H, the one-resonator specific energy-based entropy.

Environmental effect

We now examine the energy footprint attributable to environmental change along the elevation gradient set by the height of the benches. In an active floodplain such a gradient controls the frequency and type of floods, and through these the quality of the local environment. This control is independent from the phylogenetic effect in time and it is measured differently. We base all calculations on 40 taxa and 45 sample plots summarised for benches in in Table 9.

By summing (T) and counting the elements (n), we have the basic data to parametrise nH:

	B1	B2	B3	Env
T	1094	1550	828	3472
n	14	20	11	3
nH	70.6725	111.2034	57.8823	24.1629
H	5.0480	5.5602	5.2620	8.0543

In these terms, the energy footprint of the bench-height related effects is 24.2 nats for nH and 8.1 nats for H. The reader may ask at this point, why do we use n=3 in the E_{Env} column and why not 45? The reason is simple: the benches stratify the sample into 3 group of sample plots, not 45.

Overall partial effects

We base the calculation of this section's joint effect on T and n values from three record sets. The first is the original 40 taxa x 45 sample plots record set. For this the grand total is T = 3472 and the number of non-zero values n = 810. So we have for this case the specific terms of the potential energy equation $E=E_{Phy}+E_{Env}+E_{Rnd}$:

Case 1	40x45	T=3472	n=810		
Source	nH	%	n	H	P
Phylogeny	218.7738	10.5	40	5.4693	0.0042
Environment	24.1629	1.2	3	8.0543	0.0003
Joint (total)	2076.806	100.0	810	2.5639	0.0770
Grab-bag	1833.869	88.3	767	2.3955	0.0915

Two more include the records for the 40 taxa by 36 sample plots set and the 73 taxa by 45 sample plots set:

Case 2	40x36	T=2784	n=661		
Source	nH	%	n	H	P
Phylogeny	209.9966	12.4675	40.0000	5.2499	0.0052
Environment	23.5007	1.3952	3.0000	7.8336	0.0004
Joint (total)	1684.3515	100.0000	661.0000	2.5482	0.0782
Grab-bag	1450.8543	86.1373	618.0000	2.3477	0.0956

Case 3	73x45	T=3723	n=941		
Source	nH	%	n	H	P
Phylogeny	360.7343	15.3818	73.0000	4.9416	0.0071
Environment	24.3722	1.0392	3.0000	8.1241	0.0003
Joint (total)	2345.1996	100.0000	941.0000	2.4922	0.0827
Grab-bag	1960.0930	83.5789	865.0000	2.2660	0.1037

Based on the estimated percentages, we can partition any nH value of Table 4 in proportion s as given below:

Proportions				
Source	40x45	40x36	73x45	Average
Phylogeny	0.1050	0.1247	0.1538	0.1278
Environment	0.0120	0.0140	0.0104	0.0121
Joint (total)	1.0000	1.0000	1.0000	1.0000
Grab-bag	0.8830	0.8614	0.8358	0.8601

The elementary $E=E_{Phy}+E_{Env}+E_{Rnd}$ partitions are based on the proportions in the Average column :

nH

Position	R1	R2	R3
1	45.401	45.401	40.181
2	44.826	39.803	42.975
3	50.699	53.201	46.738
4	42.975	42.31	48.292
5	54.95	53.629	41.139
6	44.812	44.826	51.862
7	51.75	38.028	43.615
8	48.41	47.314	56.636
9	48.654	43.615	50.919
10	54.593	46.738	44.231
11	43.156	45.258	48.622
12	45.907	44.826	45.693

E_{Phy}

Position	R1	R2	R3
1	5.8022	5.8022	5.1351
2	5.7288	5.0868	5.4922
3	6.4793	6.7991	5.9731
4	5.4922	5.4072	6.1717
5	7.0226	6.8538	5.2576
6	5.7270	5.7288	6.6280
7	6.6137	4.8600	5.5740
8	6.1868	6.0467	7.2381
9	6.2180	5.5740	6.5074
10	6.9770	5.9731	5.6527
11	5.5153	5.7840	6.2139
12	5.8669	5.7288	5.8396

Eenv

Position	R1	R2	R3
1	0.5494	0.5494	0.4862
2	0.5424	0.4816	0.5200
3	0.6135	0.6437	0.5655
4	0.5200	0.5120	0.5843

5	0.6649	0.6489	0.4978
6	0.5422	0.5424	0.6275
7	0.6262	0.4601	0.5277
8	0.5858	0.5725	0.6853
9	0.5887	0.5277	0.6161
10	0.6606	0.5655	0.5352
11	0.5222	0.5476	0.5883
12	0.5555	0.5424	0.5529

Grab-bag

Position	R1	R2	R3
1	39.0494	39.0494	34.5597
2	38.5548	34.2346	36.9628
3	43.6062	45.7582	40.1994
4	36.9628	36.3908	41.5359
5	47.2625	46.1263	35.3837
6	38.5428	38.5548	44.6065
7	44.5102	32.7079	37.5133
8	41.6374	40.6948	48.7126
9	41.8473	37.5133	43.7954
10	46.9554	40.1994	38.0431
11	37.1185	38.9264	41.8198
12	39.4846	38.5548	39.3005

Its important to note that these are expectations as much as the proportions are too expectations under the assumption of homogeneity as in the normal spectrum.

This is how the energy structure appeared overall on the belt transect in 1976. An interesting trend is verified. The contribution of phylogeny tops the contribution of the environmental effect by one order of magnitude. A refined environmental classification, meaning an increase of the categories from 3 to higher would make an improvement in the E_{Env}'s share. ,The share of the grab-bag is consistent with the usual very heavy load of random variation found scooped up by vegetation data.

In closing

Our results have a finality we did not intend. The invasive acts by the Municipality of Hope in the Coquihalla floodplain site after the completion of our vegetation survey in 1976 have erected a barrier of isolation between past and present. Everything changed. The site is no longer an active, natural floodplain.

We are, however, not prevented from posing question and drawing general conclusions to sharpen the reader's image of analysis and results:

Q1. What kind of energy did we map? We mapped energy-based entropy, a proxy or alternative parameter of the potential energy. We did not touch calorific flow.

Q2. We parametrised the potential energy equation by T and n. What does this imply? It implies that the analysis is stand (complex) level, and the approach is holistic, *i.e.* the analysis does not penetrate down to the level of taxa (resonators).

Q3. Are there regularity conditions that have to be fulfilled to justify the use of the potential energy equation? Yes, there are such conditions. The most fundamental condition requires the assumption that the assembly of taxa into a stand-level community is a random process. We see this fulfilled within a sample plots which is free of pattern, environmental and compositional, other than random. If resources and time permit, statistical homogeneity tests can be performed.[6]

Q4. Can we consider a community assembly rule natural which assumes a random assembly process? This should not be a far fletched assumption for areally small vegetation stands.

Q5. What does an increasing nH value from the Class-level to the Species-level imply in evolutionary terms? It is telling us that the evolutionary proliferation of taxa feeds potential energy into the plant community.

Q6. If energy is transmitted through the channels of inheritance at the nods of the phylogenetic tree, how much is transmitted? The dnH value used in the section on phylogeny is a proxy measure of potential energy. The amount is variable and obviously stand specific.

Q7. Taken all into account, and considering the data-based context of the conclusions, what do we see of a general nature emerging in nH terms? There are several such things:

a. The phylogenetic effect overwhelms the current environmental effect. Both are far outweighed by the random effects.

b. The potential energy footprint of the community complex is loaded with effects from inherent stochasticity, sampling error and measurement error. We can reduce the sampling error and measurement error at a potentially substantial cost. The effect of stochasticity remains as Nature hands it to us.

---

[6] For key concepts and techniques readers should refer to
http://www.amazon.ca/Handbook-Spatial-Point-Pattern-Analysis-Ecology/dp/142008254X
Greig-Smith (1983) and Ripley (1981) describe effective pattern analytical techniques on a broad scale of mathematical complexity.

c. We addressed energy structural instability/stability in the stand-level plant community and concluded that instability increases with increasing taxon richness. A species rich stand is more likely to flip into one of its ghost energy structural states by chance than a species poor stand.

Q8. Is energy in our usage a metaphor? No, it is not. To explain this we tart with the observation that energy has never been defined, therefor it cannot be directly measure. In this sense scaling the energy level under any circumstances involves measurement of its manifestation. Therefore the unit of measure is specific to the king of manifestation. In the present case the scale used by $E = - \ln P$ renders a proxy expression of the energy level in natural units (nats). Let us recapitulate the scaling process. We started with measurement of resonator (taxon) performance on some discrete scale. We have defined each measurement as an energy unit count, following this we applied Max Planck's energy-based entropy function. We ended up with a proxy expression of the energy level of the complex (vegetation stand). If we started with air free water and wanted to know the amount of energy needed to raise the temperature of the water from 287.65 to 288.65 ^{O}K we would need to know the amount of water m in grams and apply the scalar function $Cal = 4.1855 \times m$ joules. Clearly, James Prescott Joule's scale is just another proxy scale for energy among as many proxy scales as there are different manifestations of spent energy.

Epilogue

I express my fervent hope that the Essay has succeeded in presenting the energy-based entropy option as a sound, holistic approach for the statistical study of plant community energetics. It is offered not as a replacement of the classical calorific approach, which entombed energetic studies far too long in my field, but as second choice. I also hope that students will exploit the numerous possibilities left open for further development of Statistical Quantum Ecology.

References

Braun-Blanquet, J. 1928. Pflanzensoziologie. Grundzüge der Vegetationskunde. Biol. Studienbücher, 7. 330 p. Berlin.[7]

Brillouin, L. 1962. Science and information theory. 2nd ed. Academic Press, New York.

Greig-Smith, P. 1983. Quantitative plant ecology. 3rd ed. Blackwell Scientific, London.

Kullback, S. 1959. Information theory and statistics. Wily, New York. [Revised Dover edition 1968.]

McIntosh, R.P. 1967. An index of diversity and the relation of certain concepts to diversity. Ecology 48: 392-404.

Orlóci, L. 1965. The Coastal Western Hemlock Zone on the south-western British Columbia Mainland. Vegetation-environmental patterns and ecosystem classification. In: V.J.

---

[7] --1932. Plant sociology, the study of plant communities. English translation of the 1928 edition revised and edited by G.D. Fuller, University of Chicago, and H.S. Conard, Grinnell College, U.S.A.
-- 1964. Pflanzensoziologie. Grundzüge der vegetationskunde. 3rd ed. Springer, Wien-New York, 865 p

Krajina (ed.), Ecology of Western North America. Vol. 1, pp. 18-34.

Orlóci, L. 1991. On character-based plant community analysis: choice, arrangement, comparison. Coenoses 6: 103-107.

Orlóci, L. 2006. Diversity partitions in 3-way sorting: functions, Venn diagram mappings, typical additive series, and examples. Community Ecology 7: 253-259.

Orlóci, L. 2012. Statistical Ecology. The quantitative exploration of nature to reveal the unexpected. Scada Publishing, London. 369 p. Expanded and revised 2014. Online Edition: https://createspace.com/3476529

Orlóci, L. 2013a. Quantum Ecology. The energy structure and its analysis. SCADA Publishing, Canada. Online Edition: https://www.createspace.com/4406077

Orlóci, L. 2013b. Quantum analysis of primary succession. The energy structure of a vegetation chronosere in Hawai'i Volcanoes National Park. SCADA Publishing, Canada. Online Edition: https://www.cratespace.com/4452597

Orlóci, L. 2013c. On the energy structure of natural vegetation. In search for community governance rules. SCADA Publishing, London, Canada. Expanded Online Edition: https://www.createspace.com/4153484

Orlóci, L. 2014. The vegetation process. A holistic study of long-term community energetics in East Beringia. SCADA Publishing, Canada. Online edition: https://createspace.com/4760258

Pielou, E.C. 1977. Mathematical ecology. Wiley-Interscience, New York.

Planck, Max. 1901. On the law of distribution of energy in the normal spectrum. Annalen der Physik Vol. 4, p. 553 et seq.

Rényi, A. 1961. On measures of information. In: J. Neyman (ed.), Proceedings of the 4th Berkeley Symposium on mathematical statistics and probability, pp. 547-561. University of California Press, Berkeley.

Ripley, B.D. 1981. Spatial statistics. Wiley, New York.

Shannon, C.E. 1948. The mathematical theory of communication. Bell System technical Journal 27:379-423.

Index

periodicity, 42
phylogenetic effect, 64, 70
phylogenetic structure, 20
phylogenetic tree, 21, 54, 57,
   58, 59, 60, 61, 64, 70
phylogeny, 21, 25, 67
phytosociologist, 26
Pielou, 26, 74
plant community, 26, 70, 71,
   74
plant populations, 22, 23
Porto Alegre, 10
potential energy, 7, 10, 22, 23,
   24, 25, 26, 27, 29, 30, 46,
   47, 48, 51, 54, 57, 58, 59,
   61, 63, 65, 69, 70
probabilistic index, 52
probabilities, 27
proxy tree, 21, 54
Pseudotsuga menziesii, 12, 19,
   21, 55
qualitative component, 21
quantitative component, 21
quantum, 30
quantum analysis, 30
Quantum analysis, 4, 24, 30
Quantum Ecology, 4, 24, 26,
   74
Rajski's metric, 37, 38
random process, 70
Red-cedar, 12
regression analysis, 35, 42
regression graph, 44
Rényi, 23, 24, 27, 28, 30, 31,
   36, 74

residuals, 44, 63
resonator complex, 7, 22, 26
resonator level, 24
resonator probabilities, 27
resonators, 26
Ripley, 70
sample plot, 16, 22, 23
sample plots, 16
sampling design, 26
sampling error, 70
sampling errors, 70
sampling grid, 32, 39, 41, 57
sequential, 22
Shannon, 23, 26, 27, 30, 75
slope, 42, 43
species, 18, 20, 21, 22, 23, 25,
   27, 34, 39, 46, 54, 62, 71
species rich community, 46
specific dnH, 63
specifics, 63
stability, 28, 29, 39
stand, 7, 10, 22, 23, 24, 28, 38,
   39, 69, 70, 71
stand-level, 70, 71
stochasticity, 70
superposition, 47
survey site, 12
T, 28
taxonomic code, 54
taxonomic structure, 54
theoretical framework, 26
Thuja plicata, 12, 18, 19, 20,
   55
topography, 10, 14
total energy, 25, 26, 29, 63

Supplementary references

THE VEGETATION PROCESS: A holistic study of long-term community energetics in East Beringia
Authored by Dr Laszlo Orlóci

6" x 9" (15.24 x 22.86 cm)

Black & White on White paper

216 pages

ISBN-13: 978-1499142068 (CreateSpace-Assigned)

ISBN-10: 1499142064

BISAC: Science / Life Sciences / Ecology

ORDER FROM CREATESPACE E-STORE:

https://www.createspace.com/4760258

Process, as the Book uses this term, implies simultaneous execution of two fundamental functions in continuity. One creates complexity, the other reduces it. Ecologists refer to these as community assembly and disassembly. The process requires energy input which determines the momentary potential energy state of the community. This is measurably true in terms of Max Planck's energy-based entropy. We find potential energy increasing when new species (taxa, community elements) are added to or others proliferate in the community, and decreasing when species drop out or their performance declines.

Quantum analysis of primary succession: The energy structure of a vegetation chronosere in Hawaii Volcanoes National Park

ored by Laszlo Orlóci FRSC

List Price: $30.00

6" x 9" (15.24 x 22.86 cm)
Black & White on White paper
54 pages

ISBN-13: 978-1492788997 (CreateSpace-Assigned)
ISBN-10: 1492788996
BISAC: Science / Life Sciences / Ecology

The book revisits the classical idea that the potential energy structure of primary succession is a seamless fusion of foot-prints specific to basic processes which operate on all scales – phylogeny, environmental mediation, and chance. The idea is tested in quantum analysis of a vegetation chronosere in Hawaii Volcanoes National Park. How is the test constructed? What are the outcomes? What do the results tell about primary succession not already known from other sources? Stated in the briefest of terms the test re-quires temporal species performance data...

ORDER FROM CREATESPACE ESTORE:
https://www.createspace.com/4452597

Quantum ecology: Energy structure and its analysis

Authored by László Orlóci FRSC

List Price: $30.00

6" x 9" (15.24 x 22.86 cm)
Black & White on White paper
72 pages

ISBN-13: 978-1492183297
ISBN-10: 1492183296
BISAC: Science / Life Sciences / Ecology

Ecology joins forces with quantum theory on the pages of "Quantum Ecology" to create a holistic approach in energy studies.

The infusion of quantum theoretical principles allows the study focus of ecological energetics to shift from the conventional calorific (trophic) flow in ecosystems to the potential energy structure of the vegetation. The books contents cover the theory and techniques in a unique account centred on the energy equation. The equation's component terms define energy footprints specific to ecology's basic processes, such as historic phylogeny, current environmental mediation of transience, and chance. What gives practical value to quantum analysis is its ability to be parameterised by the usual type of survey or experimental data.

The book is offered for classroom use in advanced courses and technical support in research projects.

ORDER FROM CREATESPACE ESTORE:
https://www.createspace.com/4406077

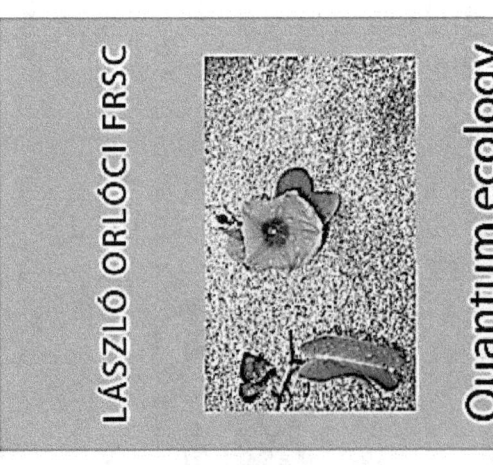

LÁSZLÓ ORLÓCI FRSC

Quantum ecology

Energy structure and its analysis

Statistical ecology

Like 0

The quantitative exploration of Nature to reveal the unexpected
Authored by Laszlo Orlóci Ph.D.

The book's topics traverse many problem areas in univariate and multivariate data analysis, focussed on current ecological practice. The manner of presentation emphasizes reasoned methodological choices and encourages innovations consistent with the objectives, but mindful of the need to see clearly the regularity conditions which set limits for valid application of statistics in Ecology. The main text is accompanied by external appendices including a technical manual, 47 specialized application programs, and many data files taken from the exercises in the main text. For information please contact: lorloci@uwo.ca

List Price: $49.90

Add to Cart

About the author:
Orlóci is an INTECOL Distinguished Statistical Ecologist. He is external (academician) Member of the Hungarian Academy of Sciences, and regular (academician) Fellow of the Academy of Sciences of the Royal Society of Canada. He published over 100 papers in scientific journals, numerous monographs and books. His current essays on trajectory analysis, the rules of process governance, and the phylogenetic signal in vegetation transitions have considerable significance for evolutionary ecology and global change science. His present work on energy structures in metacommunities is seminal, pointing to a new direction.

Publication Date:	Aug 10 2010
ISBN/EAN13:	1453760520 / 9781453760529
Page Count:	372
Binding Type:	US Trade Paper
Trim Size:	6" x 9"
Language:	English
Color:	Black and White
Related Categories:	Science / Life Sciences / Ecology

Statistical multiscaling in dynamic ecology

f Like 0

Probing the long-term vegetation process for patterns of parameter oscillation

Authored by László Orlóci Ph.D.

The Book's conceptualisation of multiscaling theory presents the Next Step in the study of the long-term vegetation process. The context is statistical and the process generating events have proxy in the compositional transitions of the palynological spectra. Familiarity with multiscaling is not a pre-requisite. The reader shall learn from the examples how multiscaling techniques helped to identify the self-similar (fractal) nature of the process, isolate low and high instability phases, locate hotspots of compositional transitions, and link these to delayed climatic effects. He or she shall also learn how to gauge process homeomorphy among sites, isolate the random and directed effects found braided into the process, and do much more within a broad yet formal probabilistic framework. The Book's contents are taken in part from a graduate course offered in the Ecology program at UFRGS in Porto Alegre, Brazil. The examples use palynological spectra from sites on the Hungarian Great Plain and in the adjacent Carpathian Mountains. Application programs are available from the author.

List Price: **$30.00**

Add to Cart

Publication Date:	Mar 15 2012
ISBN/EAN13:	1475071388 / 9781475071382
Page Count:	96
Binding Type:	US Trade Paper
Trim Size:	6" x 9"
Language:	English
Color:	Black and White
Related Categories:	Science / Life Sciences / Ecology

Self-organization and mediated transience in plant communities

Like 0

What are the rules?
Authored by Dr. László Orlóci FRSC

A novel, signal theoretical solution is sketched out for the ecological problem of how to identify and quantitatively express the assembly rules of plant communities. A case study for testing the solution leads to the astonishing conclusion that the phylogenetic signal outperforms the current environmental signal in intensity close to 7 to 1. This indicates high stability and low inclination to environment mediated transience in the community.

About the author:

László Orlóci was born in Hungary in 1932. He holds degrees in forest engineering (DFE Sopron), forestry science and biology (BSF, MSc, PhD University of British Columbia), and DSc h.c. in biology (University of Trieste). Orlóci held appointments as NATO Science Fellow (University College of North Wales), professor (University of Western Ontario), and visiting professor at universities in the Americas, the Pacific, Asia, and Europe. He is an INTECOL Distinguished Statistical Ecologists; external (academician) member of the Hungarian Academy of Sciences, and regular Fellow of the Academy of Sciences of the Royal Society of Canada.

List Price: $25.00

Add to Cart

Publication Date:	Nov 11 2011
ISBN/EAN13:	1461028221 / 9781461028222
Page Count:	70
Binding Type:	US Trade Paper
Trim Size:	6" x 9"
Language:	English
Color:	Black and White
Related Categories:	Science / Life Sciences / Ecology

On the energy structure of natural vegetation

f Like | 0

List Price: $30.00

Add to Cart

In search for community governance rules
Authored by Laszlo Orloci FRSC

Briefly about the book ...

Vegetation Science meets quantum theory in the energy-based entropy model of this book. The model is based on Max Planck's postulate that potential energy and entropy are interchangeable parameters in resonator complexes. What is a typical outcome of the model in vegetation studies? The model hands users a set of entropy estimates and probabilities based on which the strength and uniqueness of a metacommunity's energy structure can be characterised in comparative terms.

About the author:

Orlóci is an INTECOL Distinguished Statistical Ecologist. He is external (academician) Member of the Hungarian Academy of Sciences, and regular (academician) Fellow of the Academy of Sciences of the Royal Society of Canada. Orlóci published over 100 papers in scientific journals, numerous monographs, books and book chapters. His current essays on trajectory analysis, the rules of process governance, and the phylogenetic signal in vegetation transitions have considerable significance for evolutionary ecology and global change science. His present work on energy structures in metacommunities is seminal, pointing to a new direction.

Publication Date: Jan 30 2013
ISBN/EAN13: 1482319373 / 9781482319378
Page Count: 46
Binding Type: US Trade Paper
Trim Size: 6" x 9"
Language: English
Color: Black and White
Related Categories: Science / Life Sciences / Ecology

Flexible computing in statistical ecology [f]Like 0

External appendix to accompany L. Orlóci's Statistical Ecology

Authored by Dr. László Orlóci

Problem flexible computing in statistical ecology

The Book describes more than 40 executable (.exe) computer programs and presents examples of application which correspond to the examples included in Statistical Ecology*. The programs are flexibly problem specific and conversational. They allow option-driven selective access to specific statistical tasks. Linkages are possible through standard output and input. The description includes in each case a brief introduction, a record of the start up dialogue, and detailed record input and output sets. The source code is in True Basic. The programs are compiled and linked on a 32 bit Windows XP system and tested up to Windows 7.
The executable program library, data files and a note to users are distributed free of charge to purchasers of the Technical Manual. Requests for download information should be directed to the URL address lorloci@uwo.ca. Prior to running the application programs, installation of a recent version of True Basic (see Internet for sources) on the user's system is strongly advised.
* Orlóci, L. 2010. Statistical Ecology. The quantitative exploration of nature to reveal the unexpected. Scada Publishing, Online Edition. Copies are available from the distributor
https://www.createspace.com/3476529

Publication Date:	Apr 05 2011
ISBN/EAN13:	1460972953 / 9781460972953
Page Count:	142
Binding Type:	US Trade Paper
Trim Size:	6" x 9"
Language:	English
Color:	Black and White
Related Categories:	Science / Life Sciences / Ecology

List Price: $30.00

Add to Cart

Statistical Ecology. A reasoned approach.

Reader's notes

Bibliographic notes ...

László Orlóci was born into a military family in Hungary in 1932. He holds degrees in forest engineering (DFE Sopron), forestry science and biology (BSF, MSc, PhD University of British Columbia), and DSc *h.c.* in biology (University of Trieste); held appointments as NATO Science Fellow (University College of North Wales), professor (University of Western Ontario), and visiting professor at universities in the Americas, the Pacific, Asia, and Europe. He is INTECOL's Distinguished Statistical Ecologists, external academician member of the Hungarian Academy of Sciences, and regular Fellow of the Academy of Sciences of the Royal Society of Canada.

He published over 100 papers in scientific journals, numerous monographs and books. His current monographs address trajectory analysis, the rules of process governance, the phylogenetic signal in vegetation transitions, and quanrum ecology.

Orlóci is married to Márta Mihály, a Sopron forest engineering alumna. Their daughter Martha is a Geography graduate of Western University in Canada. He has two granddaughter. Kathryn Orlóci-Goodison is third year environmental resources management major and Ruth Orlóci-Goodison first year Presidential Scholar in biology, both at Lakehead University in Thunder Bay.

2015-05-09

HEADLINE ...

Quantum Ecology provides the tools needed to map the vegetation's energy structure onto a ground grid. The mapping parameters are components of the energy structure issuing from historic phylogeny, current environmental forcing, and ubiquitous chance events. Concomitant instability maps reveal increased energy structural stability under decreasing species richness.

BACKPAGE text ...

The book presents further details of a new paradigm intended for statistical studies of stand-level vegetation energetics. The approach is holistic and the techniques are quantum ecological. The case study's objective is stand-level vegetation mapping by energy criteria. Since the vegetation process is conceived as an energy structural phenomenon, the obvious choice for mapping is an energy structural component.

The book's classification model assumes a three-parted structure with components issuing from historic phylogeny, current environmental mediation, and the ubiquitous random events. The modus operandi is simple: isolate the structural components analytically, construct the maps for display on the sampling grid, then probe the maps and other numeric results for generalizable regularities. Regarding the data base, most types of vegetation survey data are admissible.

www.ingramcontent.com/pod-product-compliance
Lightning Source LLC
Chambersburg PA
CBHW070834180526
45168CB00002B/836